UNDERSTANDING ENGINEERING THERMO

UNDERSTANDING ENGINEERING THERMO

Octave Levenspiel

Chemical Engineering Department
Oregon State University

For book and bookstore information

http://www.prenhall.com

Prentice Hall PTR
Upper Saddle River, NJ 07458

Library of Congress Cataloging-in-Publication Data

Levenspiel, Octave.
    Understanding engineering thermo / Octave Levenspiel.
       p.   cm.
    Includes index.
    ISBN 0–13–531203–5
    1. Thermodynamics.   I. Title.
TJ265.L395   1996
621.402′1—dc20   96–1546
                                              CIP

Editorial/production supervision: Pine Tree Composition, Inc.
Cover design director: Jerry Votta
Cover design: Scott Weiss
Manufacturing buyer: Alexis R. Heydt
Acquisitions editor: Bernard Goodwin

© 1996 by Prentice Hall PTR
Prentice-Hall, Inc.
A Simon & Schuster Company
Upper Saddle River, NJ 07458

Printed in the United States of America
10 9 8 7 6 5 4 3 2 1

ISBN 0-13-531203-5

Prentice Hall International (UK) Limited, *London*
Prentice Hall of Australia Pty. Limited, *Sydney*
Prentice Hall Canada Inc., *Toronto*
Prentice Hall Hispanoamericana, S.A., *Mexico*
Prentice Hall of India Private Limited, *New Delhi*
Prentice Hall of Japan, Inc., *Tokyo*
Simon & Schuster Asia Pte. Ltd., *Singapore*
Editora Prentice-Hall do Brasil, Ltda., *Rio de Janeiro*

CONTENTS

16 IDEAL GASES AND THE SECOND LAW   196

17 ENTROPY OF ENGINEERING FLUIDS   208

18 WORK FROM HEAT   216

19 EXERGY OR AVAILABILITY   238

20 THERMO IN MECHANICAL ENGINEERING   256

PREFACE

My favorite volume on thermo[1] starts with a quote from Josia Willard Gibbs' cousin when she looked at a copy of his most famous paper on this subject. Her reaction was

> It looks full of hard words and signs and numbers, not very entertaining or understandable looking, and I wonder whether it will make people wiser or better.

Here's the challenge for me—to make people wiser without those hard words. Thermo's guru Kenneth Denbigh[2] wrote

> Thermodynamics is a subject which needs to be studied not once but several times over at advancing levels. In the second and third rounds . . . it is useful once again to go over the basis of the first and second laws, this time in a more logical sequence.

This book is not planned as a second or third round on the subject, but as a first introduction. Thus, I try to present thermo's main ideas and show its use with numerous practical problems.

The typical opinion of a student who first collides with the subject in a second or third round course is quoted by Andrews[3]

> To me, thermodynamics is a maze of vague quantities, symbols with superscripts, subscripts, bars, stars, circles, etc., getting changed along the way, and a dubious method of beginning with one equation and taking enough partial differentials until you end up with something new and supposedly useful.

[1]S. W. Angrist and L. G. Helper, *Order and Chaos* (New York: Basic Books, 1967).
[2]K. G. Denbigh, *The Principles of Chemical Equilibrium*, 3rd ed. (Cambridge University Press, 1971).
[3]F. C. Andrews *Thermodynamics: Principles and Applications* (New York: 1971) Wiley-Interscience.

I have sympathy for that student. Where practical I avoid the logical sophistication and abstract aspects of the subject that are easier to digest on the second or third round. As I see it, the study of thermo should be useful, interesting, and fun—not a chore. Also, I hope the reader will excuse me for using the common term "thermo" instead of the complete "thermodynamics."

As for the content of this book, the first nineteen chapters should be a reasonable foundation for a first course on the subject, while the last six should whet the appetite of the student for more advanced study.

To Peggy Blair and Marta Follett, my gratitude for their patience in redoing version after version of my handwritten manuscript. And finally, I am particularly indebted to my Mechanical Engineering colleague, Professor Murty Kanury, who used our numerous hot Korean lunches to discuss and argue a number of sticky points on this subject.

NOTATION

Symbols and constants that are defined and used locally are not listed here.

c	mass concentration, kg/m^3
c_p	specific heat at constant pressure, $J/mol \cdot K$; see Eq. 8–2
c_v	specific heat at constant volume, $J/mol \cdot K$; see Eq. 8–1
C	molar concentration, mol/m^3; see Eq. 2–1
C	$= 3 \times 10^8$ m/s, speed of light; see Eq. 6–8
e	total energy per unit mass; J/kg
$\dot{e}$	rate of change of total energy per unit of a batch system, $J/kg \cdot s = W/kg$; see Eq. 3–4
E	total energy of system, J; see Eq. 3–1
$\dot{E}$	rate of change of total energy of a batch system, $J/s = W$; see Eq. 3–3
E_k	kinetic energy, J; see before Eq. 3–1
E_p	potential energy, J; see before Eq. 3–1
g	acceleration of gravity, m/s^2; see Eq. 5–1
g	Gibbs free energy per mole of system, J/mol
g_c	conversion factor, needed when using other than SI units, see Eq. 5–1
G	moles or molar flow rate of gas, mol or mol/s; see Chapter 21
G	Gibbs free energy, J; see Eq. 22–7
G	Newton's universal gravitation constant, Eq. 5–3

h	enthalpy per unit of system, J/kg or J/mol; see before Eq. 7–7
H	enthalpy, J
$\Delta H_{s\ell}$, $\Delta H_{\ell g}$	latent heat, J; see Eq. 7–1
ΔH_c, ΔH_f, ΔH_r	enthalpy change of combustion, formation and reaction for a given chemical change, J; see Chapter 9
I	information, (–); see Eq. 24–1
k	= c_p/c_v, specific heat ratio; see Eq. 11–15
k	= 1.38×10^{-23} J/molecule·K, Boltzmann constant; see before Eq. 24–8
K	chemical equilibrium constant, (–); see Eq. 23–4
K_i	phase equilibrium constant, (–); see Chapter 21
L	moles or molar flow rate of liquid, mol or mol/s; see Chapter 21
m	mass, kg; see Eq. 2–1
ṁ	mass flow rate, kg/s; see Eq. 3–5
$\overline{mw}$	molecular mass or molecular weight, kg/mol; see Eq. 2–1
n	number of moles, mol; see Eq. 2–1
ṅ	molar flow rate, mol/s
p	pressure, Pa = N/m^2; see Eq. 2–1
P	vapor pressure of pure liquid, Pa; see Chapter 21
q	heat added to unit of system, J/kg or J/mol
q̇	heat input rate per unit of batch system, W/kg or W/mol; see Eq. 3–4
Q	heat added, J; see before Eq. 3–1
Q̇	heat input rate to system, W; see Eq. 3–3
Q	an enormously large unit of energy, 6
R	= 8.314 J/mol·K, ideal gas constant; see after Eq. 2–1
s	entropy per unit of a system, J/kg·K or J/mol·K; see Chapter 15
S	entropy, J/kg or J/mol; see Eq. 15–2
t	time, s
T	temperature, K, °C, °R, °F; see Eq. 2–1
u	internal energy per unit of system, J/kg or J/mol
u̇	rate of increase in internal energy of unit of batch system, J/kg or J/mol
U	internal energy, J; see Eq. 3–1
U̇	rate of increase of internal energy of a batch system, J/s = W
v	volume per unit of system, m^3/kg or m^3/mol
v	velocity, m/s; see Eq. 6–2

V	volume, m^3; see Eq. 2–1
w	work done by unit of system, J/kg or J/mol
$\dot{w}$	power produced by unit of system, W/kg or W/mol; see Eq. 3–4
W	work done by system, J; see before Eq. 3–1
$\dot{W}$	power produced by system, J/s = W; see Eq. 3–3
x	distance, m; see Eq. 4–1
x	mass fraction or mole fraction, (–), see equation before Eq. 12–6
y_i	mole fraction of i in a gas mixture, (–); see Chapter 21
z	height above a reference level, m; see Eq. 5–2
z	compressibility factor, (–); see Eq. 12–11

GREEK LETTERS

γ	constant for polytropic processes; see Eq. 11–19
π	total pressure, Pa; see Eq. 12–2
ρ	density, kg/m^3; see Eq. 2–1
σ	surface tension, N/m; see Eq. 4–5
η	efficiency, (–); see Eq. 18–1

SUBSCRIPTS

ex	exergy
ℓg	refers to a liquid-to-gas phase change
rev	reversible process
sℓ	refers to a solid-to-liquid change

SUPERSCRIPTS

o	refers to the reference state, usually at 1 atm

UNDERSTANDING ENGINEERING THERMO

CHAPTER 1

WHAT THERMO'S ALL ABOUT

Why is slavery not fashionable today? All sorts of reasons, but primarily because it does not pay. Oxen, water buffalo, camels, and especially horses could move people and goods from here to there more efficiently. And then in the late 1700s inventors came up with a great series of concepts and devices that were responsible for the industrial revolution in Europe. They discovered how to make fire and steam do work for them. This in turn replaced animal power.

Imagine how the world changed. Instead of four horses pulling a double-decker city omnibus, instead of relying on the wind or oarsmen to push ships across the ocean, instead of using bucket and rope to lift water out of a mine, engines now existed that could do all these jobs at much lower cost.

This great revolution, the *industrial revolution*, gave birth to the steam engine, the internal combustion engine, the electric motor, the jet engine, and many other types of engines. And the practitioners, those who developed and tended these engines, were called *engineers*.

In these developments the "ability to do work" of steam, of coal, of wood was a prime consideration. How much wood would do the work of a bucket of coal or a gallon of gasoline? The term *energy* was coined[1] to mean "ability to do work." If something had lots of energy it could do lots of work. Also, energy seemed to have different faces, such as potential energy (clock weight), kinetic energy, chemical energy, and so on.

Today the machines of society are so efficient that if you wanted to transport about 1000 thermo texts such as you are now reading from here to there, 160 km away, and if you do this by physical labor, carrying them on your back, your

---

[1]It is hard for us to believe that the term "energy" was only introduced into our language by Thomas Young in 1805 in his Bakerian Lecture to the Royal Society. See D. W. Theobald, *The Concept of Energy*, (New York: Barnes and Noble, 1966).

labor would be worth less than 2¢/hour, which is a bit less than the minimum wage. Yes, machines have changed our lives irreversibly.

Thermodynamics was then developed to study energy—how much of it is in coal, wood, running water, in steam at low and high pressure, at low and high temperature, in eggs and bacon, and in candy bars. This is the concern of the *first law of thermodynamics.*

It was discovered early that although so much of this form of energy was equivalent to so much of that form of energy, you couldn't always transform all of one kind to the other. In particular, you couldn't squeeze all the energy in steam to give work. In steam locomotives a fraction of the energy in coal is always wasted. How efficiently can energy be converted from one form to another? Sadi Carnot, a brilliant youngster in the French military, answered this question, leading to the *second law of thermodynamics.*

So there it is. The first and second laws are the main business of thermodynamics. Let us say a few words about these two laws.

The first law deals with energy interchange, or how much of this kind of energy is equivalent to that kind of energy. However, it was found that one cannot always make that exchange in practice.

The second law is probably the most fascinating law in all of science, and Carnot only touched one aspect of this law. It intrudes into all sorts of areas.

- It tells what changes are possible, what are not.
- It shoots down perpetual motion machines.
- It tells the direction of time. For example, if you watch a motion picture, how can you tell whether it is being played forward or backward? Only when you see an instance of the second law. No second law, no telling the direction of time.
- Information theory, statistical mechanics, people getting older, the operation of the brain, all those have to do with the second law.

When the first and second laws are combined they lead to relations that tell how much work is available in a given situation, a concept called *availability*, or *exergy.* For example, if you have a stream of water running down a mountainside, you could put in a dam and generate electricity. But from the same amount of water in a little lake in the plains of Kansas you couldn't extract as much energy. Thus availability concerns the extractable energy of *a system in particular surroundings.* You must know both the system and surroundings to be able to tell what fraction of the total energy of a system can be extracted to do useful work.

The outline of this book is as follows:

Chapters 3–14   deal with the first law
Chapters 15–18   deal with the second law
Chapter 19   deals with exergy or availability
Chapters 20–25   deal with various applications

Our view of the world we live in is in large measure determined by the language we use to describe it. Think of that. And here thermo plays a large part. It also gives understanding to the changes of the industrial age, the age of machines. Today we are in the midst of a new revolution, the information revolution. What is the most information that a floppy disk can hold, how much can we miniaturize chips, at what rate can an optical fiber transmit information? Thermo comes in here to give limits. So anyone dealing with energy, its transformation from one form into other forms, or concerned with what changes are possible or not, has to understand the concepts of thermo. This is why all engineers have to study it, at least its fundamental concepts.

The first organization of civilian engineers was formed in England at the time of the industrial revolution, in 1811. This was the Institute of Civil Engineers. The first sentence of its constitution laid out its goals clearly and precisely, as follows:

. . . to harness the powers and forces of nature for the benefit of mankind . . .

We should reflect on this thought. It serves equally well today for the engineering profession, as it did at that time.

Finally, R. Hazen and J. Trefil listed "Science's Top 20 Greatest Hits" (see R. Pool, *Science*, *251*, 266–267 (1991)). Where then does thermo fit in the twenty great ideas of science, the most important and fundamental ideas underlying all of the sciences? Here is the beginning of the list.

1. The universe is regular and predictable.
2. One set of laws describe all motion.
3. Energy plus mass is conserved (first law).
4. Energy always goes from more useful to less useful forms (second law).

So right behind Newton's laws we have the laws of thermo.

Albert Einstein, musing over which laws of science should be ranked as supreme, put it differently. He concluded with the following observation:

A theory is more impressive the greater is the simplicity of its premises, the more different are the kinds of things it relates and the more extended its range of applicability. Therefore, the deep impression which classical thermodynamics made on me. It is the only physical theory of universal content which I am convinced that within the framework of applicability of its basic concepts will never be overthrown.[2]

PROBLEM

1. This chapter mentioned that your physical labor in carrying stuff from here to there was worth less than 2¢/hr when compared with what machines can do. Let's see if this statement makes sense. For this let us suppose that we want to transport 1000 of your thermo texts 160 km.

   (a) One way of doing this: Rent a pickup truck for the day, pack it with books, drive 160 km, unload the books, have a leisurely lunch, drive back home, refill the gas tank, and then return the pickup to the rental agency. Estimate the cost of doing this.

   (b) A second way of doing this does not use modern machines, so let's load you with say about 20 kg of books, then have you walk 160 km, deliver, and return. How long would this take you? Repeat until all the books are delivered and then calculate how much your time is worth when compared with method (a). Ignore the cost of food and shoes.

   Note: This problem suggests that it was technology that abolished slavery. This 2¢/hr is why you go to university, so that you will use your brain and the knowledge of science and engineering, rather than physical labor, for your bread and butter.

---

[2]A. Einstein, "Autobiographical Notes," in P. A. Schilpp (Ed.), *Albert Einstein: Philosopher-Scientist*, (Evanston, IL: Library of Living Philosophers, 1949).

CHAPTER 2
PRELIMINARIES

A. SYSTEM OF UNITS

All sorts of measurement units were developed and used in societies around the world, and this led to headaches as one went from one society to another. As an example, do you know that we have an English inch, an American inch, and a Canadian inch, all a little bit different one from another? Did you know that King George III of England decided that the gallon should be the volume of his chamber pot? This became the imperial gallon. He then sent his wife's chamber pot to the colonies to be the standard there. This became the U.S. gallon. Today, to stop confusion and to develop a universal measurement language for scientists everywhere, an international system of units was developed, called the SI system, or "Système Internationale."[1]

In thermo we deal primarily with work, heat, energy, and power; Tables 2–1 and 2–2 show the relationship between the different measures of these quantities. These tables and tables of other physical quantities, such as length, mass, volume, and so on, are all given at the end of this book.

EXAMPLE 2–1. Conversion of units

My automobile is powered by a 6-cylinder 200 hp engine. How many kilowatts does this translate to?

Solution

From the above tables we have

$$200 \text{ hp} \left(\frac{10^3 \text{ kW}}{1341 \text{ hp}} \right) = \underline{\underline{149 \text{ kW}}} \quad \longleftarrow$$

---

[1]Some in the United States still reject the SI system and ask: "Why change to metric now? I'm loyal to the inch. If I see a centimeter I'll step on it."

Table 2–1. Work, Heat, Energy: The SI standard is the joule (1 J = 1 N·m).

6.272 74 × 10^{24}	10^{13}	10^6
electron-volt	erg (= 1 dyne·cm)	J (joule = N·m)
This represents the work needed to push one electron "uphill" one volt.	This used to be the old metric standard. However, this is a very small unit. It's the energy needed to lift one hungry mosquito 1 cm into the air.	This is our standard of energy today. It's the energy needed to lift a cube of butter one meter into the air.

737 562	239 006	101 972	9869.233
ft·lb$_f$ (foot pound force)	cal (calorie)	kg$_f$·m	lit·atm
	The energy needed to heat one gram of water 1°C.		

947.817	239.006	0.372 506	0.277 778
Btu (British thermal unit)	kcal (a kilocalorie)	hp·hr (horsepower·hour)	kW·hr (kilowatt·hour)
This is the energy needed to heat 1 pound of water 1 degree Fahrenheit.	In food science this is called the Calorie, or big calorie.	One horsepower spent for a whole hour.	Measures electrical work. 1 kW·hr costs about 10¢ today.

27.2 × 10^{-3}	9.478 × 10^{-3}	2.39 × 10^{-5}	9.478 × 10^{-16}
m^3 natural gas	therm	ton of oil	Q
	Commonly used in the American gas industry. 1 therm = 10^5 Btu		This is a very very large unit of energy. Q = 10^{18} Btu

Table 2–2. Power, or rate of doing work. The SI standard is the watt (1 W = 1 J/s = 1 N·m/s)

10^6	1340.405	1340.483	1341.022
W	hp (hydraulic)	hp (electric)	hp
			(1 hp = 550 ft·lb$_f$/s) James Watt defined the horsepower as a horse's average power output for a full working day. The peak power output of a horse is 12~15 hp.

10^3	284.345	1	10^{-3}
kW	ton of refrigeration	MW	GW
Portable heaters, electric coffee pots, hair dryers, toasters all use 1 to 1.5 kW of electricity costing about 10–15¢/hr of operation	Energy absorbed when one short ton (2000 lb) of ice melts daily. Exactly 12 000 Btu/hr.		One big electric power station produces electric power at this rate.

EXAMPLE 2–2. Food energy

I live on about 2000 Cal of food per day. What rate of energy use does this represent?

Solution

Note that we are talking about 2000 food calories, or 2000 big calories. So from the above tables

$$\left(\frac{2000 \text{ Cal}}{\text{day}}\right)\left(\frac{1 \text{ kcal}}{1 \text{ Cal}}\right)\left(\frac{10^6 \text{ J}}{239.006 \text{ kcal}}\right)\left(\frac{1 \text{ day}}{24 \times 3600 \text{ s}}\right) = 96.85 \ \frac{\text{J}}{\text{s}} = \underline{\underline{96.85 \text{ W}}} \longleftarrow$$

This means that we live, love, and generate heat at the rate of a 100 W light bulb.

EXAMPLE 2–3. Buying soft drinks

In last month's advertisement, a 6-pack of 12 fluid ounce cans of Coca Cola cost \$1.99 and a 2-liter bottle cost \$1.29. Which is cheaper?

Note. Four different definitions of ounces are in use today

the Troy ounce	= 31.10 gm H_2O	... a mass measure
the avoirdupois ounce	= 28.35 gm H_2O	... a mass measure
the British fluid ounce	= 28.41 ml	... a volume measure
the U.S. fluid ounce	= 29.57 ml	... a volume measure

Are you confused? Use SI.

Solution

As a basis, let us calculate the cost of 1 m³ of Coke. Then for the 2-liter bottle the cost is

$$1 \text{ m}^3 \left(\frac{1000 \text{ lit}}{1 \text{ m}^3}\right)\left(\frac{\$1.29}{2 \text{ lit}}\right) = \$1290$$

For the 6-pack using the conversion at the back of the book, the cost is

$$1 \text{ m}^3 \left(\frac{33 \ 818 \text{ fluid oz}}{1 \text{ m}^3}\right)\left(\frac{1 \text{ 6-pack}}{6 \times 12 \text{ fluid oz}}\right)\left(\frac{\$1.99}{1 \text{ 6-pack}}\right) = \$935$$

So the cost ratio is

$$\left(\frac{\text{6-pack}}{\text{2-lit bottle}}\right) = \frac{935}{1290} = 0.72$$

$\underline{\underline{\text{Buy the 6-pack}}} \longleftarrow$

B.   MOLECULAR WEIGHTS AND MOLES

More properly we should talk of molar mass, not molecular weight; however, common usage lets us use these terms interchangeably. Also, we are not concerned with that furry little burrowing creature, the mole. Let's see what we are concerned with.

First of all, atoms and molecules are so small it is not convenient to talk of the mass of single particles. For example, an atom of hydrogen has a mass of

$$0.000\ 000\ 000\ 000\ 000\ 000\ 000\ 001\ 660\ \text{gm}$$

So, for convenience, we have chosen to find the mass of a bundle of 6.023×10^{23} entities, atoms, or molecules. We call this quantity *a mole* of material. Thus, it so happens that

a mole of hydrogen atoms has a mass of 1 gm

We call this quantity the *atomic weight* of the hydrogen atom. Table 2–3 gives the atomic weights of a number of the more common elements.

TABLE 2–3.   **Short Table of Atomic Weights**

Element	**Symbol**	**Atomic Weight (gm)**
Aluminum	Al	27.0
Argon	A	39.9
Calcium	Ca	40.1
Carbon	C	12.0
Chlorine	Cl	35.5
Fluorine	F	19.0
Helium	He	4.0
Hydrogen	H	1.0
Lead	Pb	207.2
Magnesium	Mg	24.3
Nitrogen	N	14.0
Oxygen	O	16.0
Phosphorus	P	31.0
Potassium	K	39.1
Silicon	Si	28.1
Sodium	Na	23.0
Sulfur	S	32.1
Uranium	U	238.1

This is the weight of 6.023×10^{23}
atoms of the element listed

Similarly, the mass of a mole of molecules is called the molecular weight, $\overline{mw}$, of that compound. For example, common sugar, called sucrose, is a combination of 45 atoms, expressed in symbols as $C_{12}H_{22}O_{11}$. Its molecular weight is then

$$\overline{mw}_{sucrose} = (12 \times 12.0 \text{ gm/mol}) + (22 \times 1.0 \text{ gm/mol}) + (11 \times 16.0 \text{ gm/mol})$$
$$= 342 \text{ gm/mol} = 0.342 \text{ kg/mol}$$

Here are a few molecular weights:

for hydrogen gas	H_2	:	$\overline{mw}$ =	2	gm/mol	= 0.002 kg/mol
for oxygen gas	O_2	:		32	gm/mol	= 0.032 kg/mol
for ozone	O_3	:		48	gm/mol	= 0.048 kg/mol
for air (21% O_2, 78% N_2, etc.):				28.9	gm/mol	= 0.0289 kg/mol
for carbon dioxide	CO_2	:		44	gm/mol	= 0.044 kg/mol
for water	H_2O	:		18	gm/mol	= 0.018 kg/mol

C. PROPERTIES OF PURE SUBSTANCES

Thermo deals with materials—solids, liquids, and gases—and shows that the clever manipulation of these materials is how we transform less useful forms of energy to more useful forms, such as coal to gasoline to power cars and lawnmowers, garbage to air condition your home, pig manure to electricity to run your TV set. So before we start on thermo let us focus on some simple properties of materials, such as pressure p, temperature T, and density ρ, and see what relationships exist between these properties.

1. Solids and Liquids

There are no nice simple equations to predict the density of liquids and solids. So densities are tabulated in various texts and handbooks.

Density of solids and liquids change but slightly with temperature (not over a factor of two) and even less with pressure. For example, for a pressure change from 1 bar to 1000 bar the density changes by 3% for liquid water, 2% for solid iron. For thermo purposes, the one value tabulated in engineering handbooks is usually sufficiently accurate. However, for a few important engineering materials, such as water, we may need more accuracy, so tabulations are given for various pressures and temperatures.

Unfortunately, there are no generalizations to predict the density of a new material. For gases things are quite different, as we will now see.

2. Gases

Many relationships between pressure, volume, and temperature have been proposed for a batch of gas, such as

$$\left.\begin{array}{l} \text{van der Waals} \\[4pt] \text{Benedict Webb Rubin} \\[4pt] \text{Beattie Bridgeman} \end{array}\right\} \text{equations,}$$

however, we like the ideal gas law because it is the very simplest and leads to all sorts of nice simple consequences and because it well approximates all real gases at not too high pressure. Since it is a good vehicle to teach thermo we will use it quite a bit in the following.

In its many equivalent forms the ideal gas (or perfect gas) law says that

$$pV = nRT = \frac{m}{mw}RT$$

$$\rho = \frac{p \cdot mw}{RT} \quad \text{or} \quad C = \frac{n}{V} = \frac{p}{RT} \qquad (2\text{--}1)$$

where volume $[m^3]$, no of moles, mass of gas $[kg]$, temperature $[K]$, pressure $[Pa = N/m^2]$, molecular mass $[kg/mol]$, gas constant $R = 8.314 \; J/mol \cdot K$, density $[kg/m^3]$, molar concentration $[mol/m^3]$.

Note that in the SI system the units and values of molecular weight are defined as

$$mw_{H_2} = 0.0020 \; kg/mol$$

$$mw_{air} = 0.0289 \; kg/mol$$

$$mw_{O_2} = 0.032 \; kg/mol$$

Other values are shown just after Table 2–3. Also, the gas constant in the SI system is

$$R = 8.314 \;\; J/mol \cdot K$$

$$= 8.314 \; Pa \cdot m^3/mol \cdot K$$

$$= 8.314 \; N \cdot m/mol \cdot K$$

Throughout this book we will use the SI system; however, any other consistent set of units can be used. Thus, the ideal gas law can be represented in other units. In that case you must use the corresponding value of the gas law constant. For example,

$$R = 1.987 \text{ cal/mol·K}$$

$$= 1.987 \text{ Btu/lbmol·°R}$$

$$= 0.082\ 06 \text{ lit·atm/mol·K}$$

$$= 0.729 \text{ ft}^3 \cdot \text{atm/lb mol·°R}$$

At not too high a pressure, say up to 5 bar, the ideal gas law reasonably represents all gases from hydrogen to uranium hexafluoride.

EXAMPLE 2–4.   A fantabulous catalyst

My research? Please keep it quiet, but I'm on the trail of an amazing powder which when sprinkled in water at exactly 30°C decomposes water into its elements, or

$$\text{water} \xrightarrow[\substack{\text{30°C, 1 bar}}]{\substack{\text{superspecial catalyst}}} \text{hydrogen and oxygen}$$

If it works (but of course it will), please tell me how many cubic meters of gas at 1 bar and 30°C can be produced by decomposing 1 liter of water. Note that 1 lit of liquid water weighs 1 kg.

Solution

In the language of chemistry the reaction proceeds as follows

$$H_2O \rightarrow H_2 + \frac{1}{2} O_2$$

or 1 mole of liquid water produces 1 mole of hydrogen gas and $1/2$ mole of oxygen gas. First calculate the number of moles of gas formed

$$(1 \text{ lit } H_2O) \left(\frac{1 \text{ kg } H_2O}{1 \text{ lit}} \right) \left(\frac{1 \text{ mol } H_2O}{0.018 \text{ kg } H_2O} \right) \left(\frac{1 \text{ mol } H_2 + \frac{1}{2} \text{ mol } O_2}{1 \text{ mol } H_2O} \right) = 83.3 \text{ mol gas}$$

So the volume of gas formed, from the ideal gas law, is given by

$$V_g = \frac{nRT}{p} = \frac{(83.3 \text{ mol}) \left(8.314 \dfrac{\text{Pa·m}^3}{\text{mol·K}} \right) (303 \text{ K})}{100\ 000 \text{ Pa}} = \underline{\underline{2.1 \text{ m}^3}} \quad \longleftarrow$$

EXAMPLE 2–5.   Combustion of waste plastic

Waste polyethylene scrap, having a chemical formula $(C_2H_4)_n$, consisting of baggies, plastic pipes, bottles, and food containers, is shredded and burned to completion with a stoichiometric[2] amount of air (Figure 2–1). Thus, the equation representing this reaction, putting n = 1 (no loss in generality), is

$$C_2H_4 + 3\,O_2 \rightarrow 2\,CO_2 + 2\,H_2O$$

(a) What volume of air entering a bit above atmospheric pressure, say at p = 103 000 Pa and 25°C, is needed to burn 3 kg of scrap?

(b) What volume of flue gas (exit gas) at 227°C and 0.9 atm leaves the burner for each 3 kg of scrap burned?

Solution

First, let us evaluate the molecular mass per unit of polyethylene. From Table 2–3 of this chapter we find that

$$mw_{C_2H_4} = [2\,(0.012) + 4\,(0.001)] = 0.028 \text{ kg/mol}$$

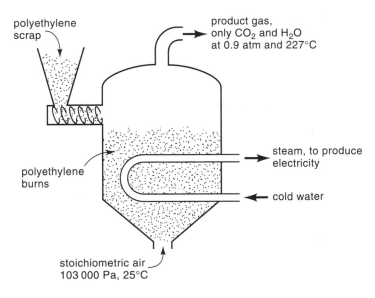

Figure 2–1.

[2]The word "stoichiometry" was coined by German chemist Richter in 1792. It refers to the art (it was an art in his time) of determining how much of this chemical would combine with a given amount of that chemical. Today chemical theory gives chemical equations to evaluate the stoichiometry of a reaction.

Now, since we are dealing with gases, not liquids or solids, it is nearly always simpler to deal in molar quantities, not masses. So let's see what this 3 kg of scrap is equivalent to in molar units.

$$(3 \text{ kg}) \left(\frac{1 \text{ mol } C_2H_4}{0.028 \text{ kg}} \right) = 107 \text{ mol of } C_2H_4$$

Next, since we have to account for many different species entering and leaving the reactor, it would be a good idea to make an accounting table to keep track of all the components. So referring to the stoichiometric equation we have, based on 107 mols of C_2H_4,

Components Involved	Entering the Reactor	Leaving the Reactor
C_2H_4 (solid)	107 mol (solid)	0
O_2	3 (107) = 321	0
N_2	(79/21) (321)	(79/21) (321)
CO_2	0	2 (107)
H_2O	0	2 (107)
Total gases	1529 mol	1636 mol

Therefore, for 3 kg of waste the volume of entering gases, just air, is

at standard conditions, 0°C and 101325 Pa

$$V_{in} = (1529 \text{ mol}) \left(\frac{0.0224 \text{ m}^3}{\text{mol}} \right) \left| \left(\frac{101\,325 \text{ Pa}}{103\,000 \text{ Pa}} \right) \left(\frac{298 \text{ K}}{273 \text{ K}} \right) \right.$$

volume at standard conditions   *pressure correction*   *temperature correction*   (a)

$$= 36.8 \text{ m}^3 \text{ air enters/3 kg waste}$$

and for the leaving combustion gases plus nitrogen

$$V_{out} = (1636 \text{ mol}) \left(\frac{0.0224 \text{ m}^3}{\text{mol}} \right) \left(\frac{101\,325 \text{ Pa}}{0.9 \times 101\,325 \text{ Pa}} \right) \left(\frac{227 + 273 \text{ K}}{273} \right)$$

$$= 74.6 \text{ m}^3 \text{ leaving/3 kg waste}$$ (b)

PROBLEMS

1. To simulate toxic material, suppose that one 42-gal barrel full of water is dumped into the ocean and is then well mixed with all the waters of all the oceans of the earth.

   If I go to Newport after this is done and scoop up one teaspoonful of water from the ocean, what is the probability of finding one of the original molecules, or how many of the original molecules may I expect to find in it?

   Data • The distance from the north pole to the equator is about 10 million meters. That is how the length of the meter was chosen.
   - Two-thirds of the earth's surface consists of water.
   - The average depth of the oceans is about 3000 m.
   - 1 mole of any substance contains 6.023×10^{23} molecules.

2. If all the human beings on our planet (roughly 4.5 American billion, or 0.0045 British billion, or better still, 4.5×10^9 men, women, and children) were all gently compacted into a cube, how large on a side would this meaty lot be?

   Note: As is often the case in real-world problems, either too much data, or contradictory data, or not enough data is given. When not enough data is given, use your good judgment and make reasonable estimates for what is unknown, but is needed.

3. *Fun at Home* is a little book that suggests creative projects for entertaining bored precocious children. Here is an example. Slip some dry ice (solid CO_2) into an empty 2 liter plastic pop bottle, seal the bottle, and set it carefully in the kitchen cabinet that stores glasses, cups, and saucers. Then, for safety reasons, be sure to close the cabinet door. When the dry ice warms up to room temperature and sublimes (vaporizes), it will explode like a medium-sized hand grenade, shattering everything in the cabinet. Neat, eh?

   By the way, how many grams of dry ice must be slipped into the bottle, knowing that the bottle will explode when its internal pressure exceeds 11 bar?

4. Factory air contains 0.1 ppb (one part per U.S. billion) of a carcinogen $\overline{mw}$ = 0.290 kg/mol that accumulates in the body when breathed. When more than 1 mg (or one milligram) of this carcinogen is absorbed in the body, it triggers an irreversible change that usually leads to lung cancer. Assuming 12 breaths/min, 0.5 lit/breath, 20% absorption of carcinogen from the 20°C room air, find how long a person has to work in this factory (40 hrs/week, 48 weeks/year) to absorb the critical dose of carcinogen (1 mg).

5. How many kilograms of oxygen are required to burn 10 kg of benzene (C_6H_6) to carbon dioxide and water vapor?

6. A 0.82 m^3 tank is designed to withstand a pressure of 10 atm. The tank contains 4.2 kg of nitrogen and I slowly heat it up, starting at room temperature. At what temperature in °C will the tank rupture?

7. We wish to market oxygen in small cylinders (0.5 ft^3) each containing 1.0 lb of pure oxygen. If the cylinders may be subjected to a maximum temperature of 120°F, calculate the pressure for which they must be designed, assuming ideal gas law behavior. The manufacturer of the cylinders will be in Singapore, so please give your answers in pascals.

8. Oxygen can be prepared according to the following reaction

$$2\ KClO_3 \xrightarrow{\ \ heat\ \ } 2\ KCl + 3\ O_2 \uparrow$$

   (a) How many grams of oxygen will be produced by the decomposition of 10 gm of potassium chlorate?

   (b) How many liters at 20°C and 98 195 Pa would this oxygen occupy?

9. The complete combustion of gasoline (take it to be pure octane or, C_8H_{18}, $\rho = 700$ kg/m^3) produces water and carbon dioxide.

   If I burn a gallon of gasoline in my car and I let the exhaust gases cool down sufficiently for all the water to condense, how many gallons of liquid water do I produce to clog and drown my catalytic muffler-converter?

10. When heated, calcium carbonate, $CaCO_3$, decomposes to give calcium oxide, CaO, and carbon dioxide. How many m^3 of CO_2 at 449°C and 120 kPa are produced per ton of $CaCO_3$ decomposed?

CHAPTER 3
FIRST LAW OF THERMODYNAMICS

Observation shows that energy can't just appear from nowhere. If a system or object gains energy, then this energy must have come from outside of it. Thus

> **The First Law, or Law of Conservation of Energy:**
> Energy cannot be created or destroyed. You can only change it from one form to another, or you can only add it to the *system* (the thing we are dealing with) from the outside, which we call the *surroundings*.

The basic unit of energy in all its forms is the Joule.

Let us take an example. Suppose you wanted to heat a cup of cold water, or in thermotalk, increase its internal energy. First of all, you can't do it if you isolate the cup and water (the system) from its surroundings. You have to add energy to it from the outside. Figure 3–1 shows some ways to do it.

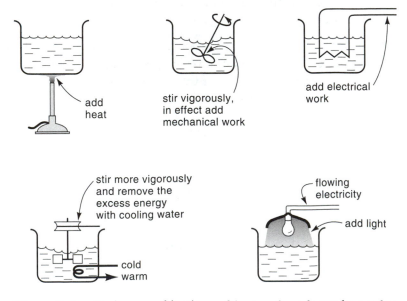

Figure 3–1. Various combinations of heat and work can be used to heat a cup of cold water.

Figure 3–2 gives us another example. Suppose I want to raise a weight.

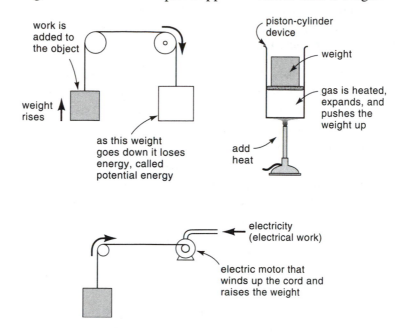

Figure 3–2.   Work and/or heat can be used to raise an object.

And again, suppose I want to speed up an object (Figure 3–3).

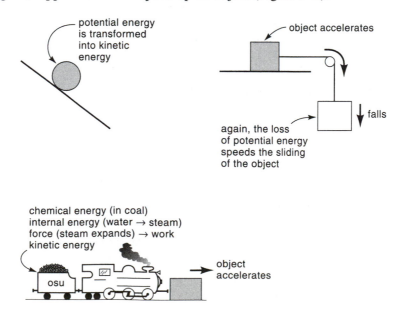

Figure 3–3.   Various forms of heat and work can be used to accelerate an object.

These examples show that adding or removing heat or work or any combination of heat and work can affect the energy of an object or system.

Let us look at the various forms of energy change of a system and the instruments that can cause this energy change.

1. **Heat**—the way of adding or removing energy from a system by contact with a hotter or cooler body. Let $+Q$ (J) be the *heat added to* the system.

2. **Work**—all ways of changing the energy of a system other than by adding or removing heat, including
   - push-pull work (piston-cylinder)
   - electric and magnetic work (electric motor)
   - chemical work (by reaction of gasoline with air in an automobile engine)
   - surface work (creating surface, making an emulsion such as mayonnaise)
   - elastic work (winding up a spring)

   Let $+W$ (J) be the *work done by* the system on the surroundings, hence $+W$ represents loss of energy by the system.

3. **Changing the Internal Energy of an Object or of a System**, ΔU, can be done in various ways:
   - by changing the temperature of a system (heat or cool)
   - by changing phase (solid to liquid, or liquid to gas)
   - by changing the molecular arrangement (chemical reaction), for example, combining carbon with oxygen as in the burning of charcoal to form carbon dioxide, or $C + O_2 \rightarrow CO_2$
   - by changing the atomic structure (nuclear fission) or breakup of large atoms to yield small fragment atoms

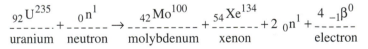

$$\underset{\text{uranium}}{_{92}U^{235}} + \underset{\text{neutron}}{_{0}n^{1}} \rightarrow \underset{\text{molybdenum}}{_{42}Mo^{100}} + \underset{\text{xenon}}{_{54}Xe^{134}} + 2\,_{0}n^{1} + \underset{\text{electron}}{_{-1}\beta^{0}}$$

   - by combining little atoms to form large atoms (nuclear fusion)

$$\underset{\text{deuterium}}{_{2}D^{1}} + \underset{\text{tritium}}{_{3}T^{1}} \rightarrow \underset{\text{helium}}{_{2}He^{4}} + \underset{\text{neutron}}{_{0}n^{1}}$$

   Let $\Delta U = U_2 - U_1$ be the change in internal energy of the system as it goes from state 1 to state 2.

4. **Changing the Potential Energy of an Object**, ΔE_p, comes from changing its location in a force field, whether gravitational, electrical, or magnetic; for example, in raising an object from the ground or on pushing an electron towards a negatively charged plate. Winding up a spring or stretching a rubber band also counts as increasing the potential energy of a system.

5. Changing the Kinetic Energy of an Object, ΔE_k, comes from changing its velocity. The faster the object moves the greater is its kinetic energy.

A. NOMENCLATURE

Consider a uniform[1] system having mass m or containing a number of moles n of material. (In some situations it is more convenient to use m, in other situations n.) Then our nomenclature is as follows.

$\mathbf{E} = \mathbf{me}$ = total energy of the system of mass m   [J]

$$\mathbf{e} = \frac{\mathbf{E}}{\mathbf{m}} \left(= \frac{\mathbf{E}}{\mathbf{n}} \right) = \text{total energy of unit mass (or of a mole) of system} \quad [\text{J/kg (or J/mol)}]$$

$$\dot{\mathbf{E}} = \mathbf{m}\dot{\mathbf{e}} = \frac{d\mathbf{E}}{dt} = \text{rate of total energy change of the system} \quad [\text{J/s or W}]$$

$$\dot{\mathbf{e}} = \frac{\dot{\mathbf{E}}}{\mathbf{m}} \left(= \frac{\dot{\mathbf{E}}}{\mathbf{n}} \right) = \frac{de}{dt} = \text{rate of total energy change per kg (or per mol) of system}$$
$$[\text{W/kg (or W/mol)}]$$

We have similar definitions for the other energy quantities that we may encounter later

$$U, E_p, E_k, H, S$$

For heat and work terms

Q, q, $\dot{Q}$, and $\dot{q}$ [J, J/kg or J/mol, W, and W/kg or W/mol] represent the various measures for *heat added to the system* and *rate of heat addition*.

W, w, $\dot{W}$, and $\dot{w}$ [J, J/kg or J/mol, W, and W/kg or W/mol] represent the various measures of *work done by the system*, and rate of doing work, called *power output*.

---

[1]Normally we take the term "uniform system" to mean one that is at constant temperature, pressure, and composition throughout. However, in the real world there is no such system. For example, even for a glass of sea water at equilibrium (at rest) the pressure at the top and at the bottom of the glass differ, and so does the composition. (If our oceans were at equilibrium with a 3% salt concentration at the surface, then in the ocean depths, about 8 km down, the concentration would be about 8%. This is not what we find, meaning that our oceans are nowhere near equilibrium.) We consider this aspect of equilibrium in Chapter 22. Still, for most systems encountered we can assume an average pressure and average composition and consider this a uniform system.

B. VARIOUS FORMS OF THE FIRST LAW

We are now ready to write the first law for various situations.

1. Isolated System

Here no mass, heat and/or work enter or leave the system in the time interval between t_1 and t_2. There may be interchange between potential, kinetic, and internal energy within the system, but not with the surroundings. Thus the total energy of the system stays unchanged, as shown in Figure 3–4. In symbols for a system of mass m we can write

$$\left.\begin{aligned} \Delta E = E_{time2} - E_{time1} = 0 \\ (U + E_p + E_k)_{time2} - (U + E_p + E_k)_{time1} = 0 \\ \Delta U + \Delta E_p + \Delta E_k = 0 \end{aligned}\right\} \text{ [J]} \tag{3–1}$$

internal   *kinetic*   *potential*

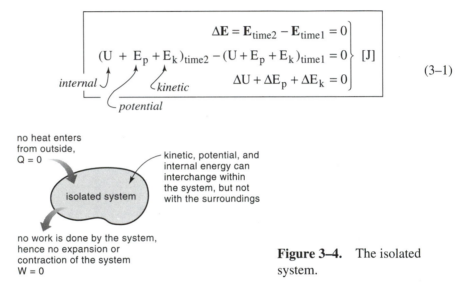

no heat enters from outside, Q = 0

kinetic, potential, and internal energy can interchange within the system, but not with the surroundings

isolated system

no work is done by the system, hence no expansion or contraction of the system W = 0

Figure 3–4. The isolated system.

2. Closed or Batch System

These "closed" or "batch" terms mean that no mass enters or leaves the system; however, heat or work can be added or removed. Thus, between time 1 and time 2 for a closed system of mass m, as shown in Figure 3–5, we may write

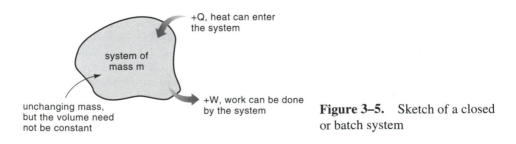

+Q, heat can enter the system

system of mass m

unchanging mass, but the volume need not be constant

+W, work can be done by the system

Figure 3–5. Sketch of a closed or batch system

$$\Delta E = \mathbf{E}_{time2} - \mathbf{E}_{time1} = Q - W$$

added to system in / *done by system in*
that time interval — *that time interval*

$$\Delta U + \Delta E_p + \Delta E_k = Q - W$$

[J]   (3–2)

In rate form for a closed system of mass m

$$\dot{\mathbf{E}} = \frac{d\mathbf{E}}{dt} = \dot{Q} - \dot{W} \quad [W]$$

(3–3)

$$\dot{e} = \frac{de}{dt} = \dot{q} - \dot{w} \quad \left[\frac{W}{kg}\right]$$

(3–4)

3. Open or Flow System

Here mass, heat, and work can all enter or leave the system, as shown in Figure 3–6. Accounting for the flow of material entering and leaving the system we write

$$\sum_{\substack{\text{streams}}} \dot{m}_{\text{outgoing}} + \dot{m}_{\text{system}} = \sum_{\substack{\text{streams}}} \dot{m}_{\text{incoming}}$$

or

$$\dot{m}_{\text{system}} \pm \sum \dot{m}_{\text{streams}} = 0 \qquad [kg/s]$$

$\begin{cases} + \textit{ is for leaving streams} \\ - \textit{ is for entering streams} \end{cases}$

(3–5)

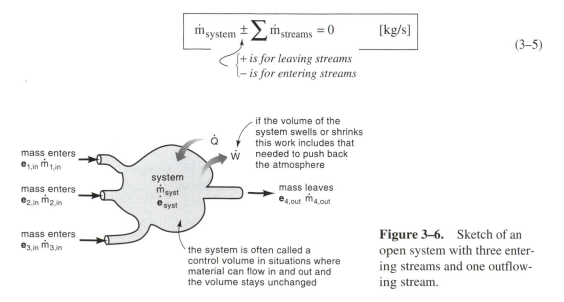

mass enters
$e_{1,in} \, \dot{m}_{1,in}$

mass enters
$e_{2,in} \, \dot{m}_{2,in}$

mass enters
$e_{3,in} \, \dot{m}_{3,in}$

system
$\dot{m}_{syst}$
$\dot{e}_{syst}$

if the volume of the system swells or shrinks this work includes that needed to push back the atmosphere

$\dot{Q}$ $\dot{W}$

mass leaves
$e_{4,out} \, \dot{m}_{4,out}$

the system is often called a control volume in situations where material can flow in and out and the volume stays unchanged

Figure 3–6. Sketch of an open system with three entering streams and one outflowing stream.

Similar to the accounting of mass, we find it more convenient and understandable to write the basic energy equation in rate form

$$\dot{E}_{system} + \underbrace{\sum \dot{E}_{\substack{outgoing \\ streams}}}_{leaving} + \dot{W} = \dot{Q} + \underbrace{\sum \dot{E}_{\substack{incoming \\ streams}}}_{entering}$$

Rearranging terms gives

$$\dot{E}_{system} \pm \sum \dot{E}_{streams} = \dot{Q} - \dot{W}, \quad [W] \tag{3-6}$$

$$\begin{cases} + \text{ is for leaving streams} \\ - \text{ is for entering streams} \end{cases}$$

$$(e\dot{m})_{system} \pm \sum (e\dot{m})_{streams} = \dot{Q} - \dot{W}, \quad [W] \tag{3-7}$$

In thermodynamic jargon we often reserve the term *system* for a closed or batch system and we reserve the term *control volume* for the open system or the section of space that we are considering.

In Chapter 13 we develop the equations for steady state flow systems, and in Chapter 14 we develop the equations for unsteady state flow systems. Until then we will deal primarily with closed or batch systems.

PROBLEMS

1. In Figure 3–1 we show four ways of heating a cup of cold water. Can you think up another different way?

2. In Figure 3–2 we show three ways of raising an object. Can you think of a still different way?

3. In Figure 3–3 we show three ways of speeding up an object. Can you think of another way?

4. Martin Mauk Smith, an Alaskan, eats, drinks, sweats, urinates, excretes solid waste including fingernails and hair, breathes in oxygen, breathes out carbon dioxide and water vapor, does hard work and his body loses heat, but his weight stays practically unchanged throughout the year.

   (a) Sketch this person (or object, or system or control volume, call him what you wish), a stick figure would do, and show the entering and leaving streams.

(b) With symbols such as $\dot{m}_{food}$ [kg/s], etc., and $\mathbf{e}_{food}$ [J/kg], write the mass balance.

(c) Write the energy balance for Martin Mauk Smith.

5. Can you think up an isolated system in which the internal energy, potential energy, and kinetic energy are all changing? Sketch such a system showing these changes.

6. A powerful electric fan is plugged in and switched on in a closed insulated room, causing the air to circulate around the room in a clockwise direction. After 14 hrs the fan is reversed and air circulates in a counterclockwise direction. After 28 hours the fan is switched off. What can you say about the temperature and energy of the room before and after the 28 hrs? Make a temperature-time sketch and an energy-time sketch for the room for these 28 hours plus an hour before and after.

7. A weight hangs by a very thin thread (ignore its mass) from a most mysterious black box that seems to be completely isolated from the surroundings. As we watch, we notice that the weight is slowly rising. What is happening to the energy of the box? Give a possible explanation of how this could happen.

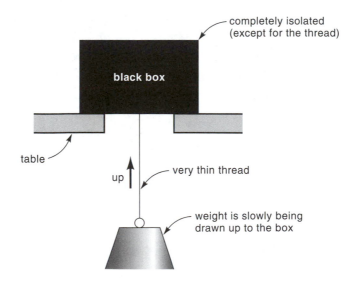

8. Squeezable but incompressible Bubbles-La-Rue is innocently floating around the swimming pool when Zzoran-the-Mean, who is skulking by the side of the pool, reaches over, pushes, and holds her under water. Naturally, our dear Bubbles is frozen with surprise. From the thermodynamic point of view, how has his dastardly act affected

(a) her energy (up, down, unchanged)?

(b) the energy of the water (up, down, unchanged)?

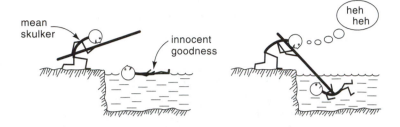

9. The electric fan in an isolated room is connected to a set of long-life high-energy batteries resting on the floor of the room. The fan is switched on and runs until the batteries are exhausted.

   (a) What can you say about the energy of the room before and after the running of the fan?

   (b) What can you say about the temperature of the room before and after the running of the fan?

10. A 10 kg mass hangs 2 m off the ground and is connected by rope and pulley to a 40 kg mass that is resting on the ground (see sketch). I reach up and with much effort manage to wrestle the 10 kg weight to the ground and hold it there. What happens to the energy of this 10 kg mass during the process? Does it go up, down, or stay unchanged?

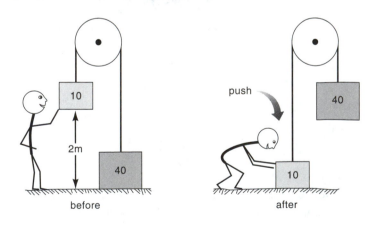

11. Two identical beakers contain the same amount of sulfuric acid of the same strength. Into the first beaker I put a tightly coiled spring; in the other I put

an identical spring, but uncoiled. The springs are attacked by the acid and dissolve. What is the difference in final state of the two beakers?

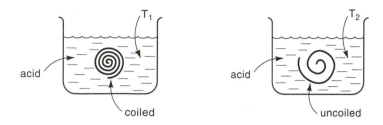

12. The door to an ordinary electric home refrigerator is left open by accident (with the power on) while the people are away for the weekend. If the kitchen doors are closed and the room is thermally well insulated, will the room be hotter than, colder than, or at the same temperature as the rest of the house when the unhappy people return? Why?

13. An insulated vertical tube half filled with water is slowly and frictionlessly rotated about its center of gravity to the horizontal position (see sketch). What happens to its temperature?

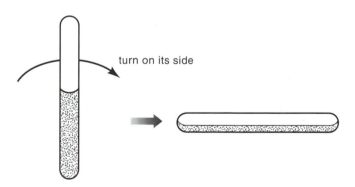

14. Two identical catapults are stretched tight and then are released. One fires a pellet, the other fires nothing. Is there any difference in final state of the rubber of the catapults?

15. From *The Feynman Lectures in Physics*, Vol. 1, pg. 45-4, California Institute of Technology (1963), we read

> ". . . let us consider a rubber band. When we stretch a rubber band we find that its temperature falls . . ."

(a) Do an experiment to check this statement. Take a rubber band, pull and release while touching the band to your lips, and then decide whether you agree with the above quote.

(b) Then try to explain your findings in terms of thermo.

16. An arrow is drawn back horizontally by a bowstring. Has the energy of the arrow gone up, down, or stayed unchanged? If you think that the energy is unchanged, why is it foolish to stand in front of the drawn arrow? (from Bob Pettit)

17. A closed system receives 100 J of heat and does 125 J of work on the surroundings.

(a) Is this possible?

(b) Give a reason for your answer.

CHAPTER 4

WORK AND HEAT

Chapter 3 introduced the three general forms of the first law—for isolated systems, for closed systems (no mass enters or leaves), and for open systems (mass can enter and/or leave). We presented the equations but did not show how the various individual heat, work, or energy terms are measured quantitatively. In this and the next five chapters we show how to do this. Thus, taking Eq. 3–2 for closed or batch systems, as illustration, we have

$$\overbrace{\Delta E_p \quad + \quad \Delta E_k \quad + \quad \Delta U}^{\Delta E} \quad = \quad \underbrace{Q - W} \tag{3–2}$$

$$\underset{Chapter\ 5}{\Big\backslash} \quad \underset{Chapter\ 6}{\Big\backslash} \quad \underset{Chapters\ 7-9}{\Big\backslash} \quad \underset{Chapter\ 4}{\Big\backslash}$$

Here we concentrate on the heat and work terms.

A. WORK

In common language the word "work" can have various meanings, such as what makes you puff, pant, and sweat; what you do from 8 AM until 5 PM; or what you are paid to do. However, science defines this word precisely,[1] with no ambiguity, and this is what we use exclusively here. However, you will see that "work" has many different faces—electrical work, chemical work, mechanical work, and so on. We will consider all these.

---

[1]The term "work" was first used in the scientific sense by Coriolis in 1829. See *Science, 173*, 118 (1971).

First, consider a system that does or receives work, and let us ignore heat effects, or put Q = 0 in the above expression.

1.   **Push-Pull Work** (Figure 4–1)

This is the basic definition of work, which was introduced and properly defined by Newton.

force applied to object

$$W = Fx$$

work done / \ *distance object moves while*
by the person   this force is applied

or more generally

$$W = \int_{x_1}^{x_2} F\,dx \qquad [\text{N·m} = \text{J}] \qquad (4\text{–}1)$$

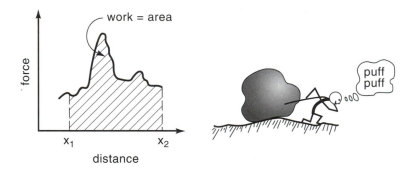

Figure 4–1.

2.   **pV Work** (Figure 4–2)

This is the useful measure for piston-cylinder processes, as in automobile engines. For a piston of cross-sectional area A, we can write

$$W = \int F\,dx = \int \frac{F}{A}\,d\,(xA)$$

force of gas pushing   *volume of gas*
on piston wall

or

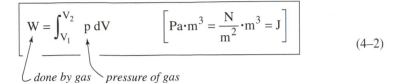

$$W = \int_{V_1}^{V_2} p \, dV \qquad \left[Pa \cdot m^3 = \frac{N}{m^2} \cdot m^3 = J \right]$$

(4–2)

└ *done by gas* ╲ *pressure of gas*

Measuring the pressure and the volume of gas in the cylinder as the piston moves tells the amount of work done by the gas.

One comment. The above equation tells the work done by the gas in the cylinder, but where does this work go?

- Some goes into pushing back the atmosphere.
- Some may go into pushing a shaft back.
- Some goes into friction and generation of heat.

However, with a reciprocating out-and-in motion of the piston, the push on the atmosphere is countered when the piston returns to its starting position. So this work term cancels and can be ignored. Also, friction is usually minor and often ignored. So as a first approximation, the work done by the gas in a piston-cylinder process is delivered to the shaft.

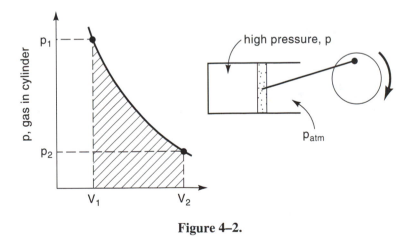

Figure 4–2.

EXAMPLE 4–1.   pV work

High pressure gas expands in a cylinder by pushing back a frictionless piston with connecting shaft (Figure 4–3), thus doing work on the shaft and the at-

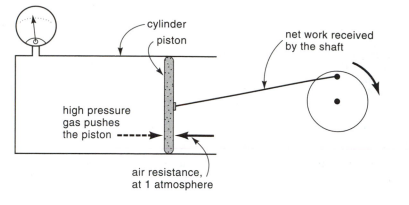

Figure 4–3.

mosphere, which is at 1 atm. During an expansion stroke the following data are recorded.

	Volume of gas in the cylinder lit	Pressure of gas in the cylinder atm		Net pressure atm
at start →	2.0	12	Outside	11
	2.4	10	air	9
	3.0	8	resists	7
	4.3	6	at	5
at end →	7.6	4	1 atm	3

How much work in joules does the gas deliver to the shaft on one expansion stroke?

Solution

Work done by the gas in one push

$$W = \int_{2.0}^{7.6} p \, dV$$

Figure 4–4 evaluates this graphically.

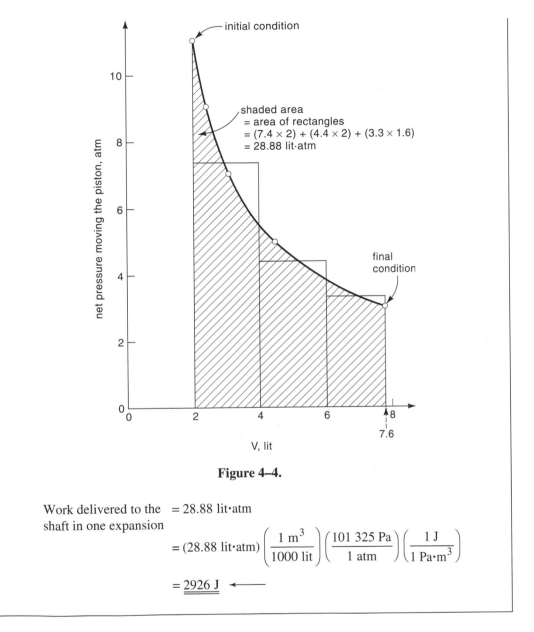

Figure 4–4.

Work delivered to the = 28.88 lit·atm
shaft in one expansion

$$= (28.88 \text{ lit·atm}) \left(\frac{1 \text{ m}^3}{1000 \text{ lit}} \right) \left(\frac{101\ 325 \text{ Pa}}{1 \text{ atm}} \right) \left(\frac{1 \text{ J}}{1 \text{ Pa·m}^3} \right)$$

$$= \underline{2926 \text{ J}} \ \longleftarrow$$

3.   Electrical Work

Let us start with some definitions. The *coulomb* is the charge, or amount of electricity, or number of electrons being pushed about, or

$$1 \text{ coulomb} = 6.24 \times 10^{18} \text{ electrons}$$

The *ampere* is the flow rate of electrons, or

$$1 \text{ ampere} = 1 \text{ coulomb/s}$$

The *volt* is the potential difference in the electrical field.
From these basic definitions we find for the flow of electricity:

$$\begin{aligned} \text{Work: } 1 \text{ J} &= 1 \text{ coulomb} \cdot \text{volt} \\ &= 1 \text{ ampere} \cdot \text{volt} \cdot \text{s} \end{aligned} \tag{4-3}$$

$$\begin{aligned} \text{Power: } 1 \text{W} &= 1 \text{ coulomb} \cdot \text{volt/s} \\ &= 1 \text{ ampere} \cdot \text{volt} \\ &= 1 \text{ J/s} \end{aligned} \tag{4-4}$$

When you pay your electric bill, you pay for the number of joules used, or about

$$10¢ \text{ per kW} \cdot \text{hr}$$

$$= 10¢ \text{ per } (1000 \text{ J/s}) \cdot (3600 \text{ s})$$

$$= 10¢ \text{ per 3.6 million joules}$$

$$= 2.8¢ \text{ per million joules}$$

So you see a joule is a rather small unit of work.

4.   Surface Tension Work

Have you ever seen an insect struggling to free itself from the surface of water? Its problem is that it has to create new surface, fresh surface, to take its place as it leaves. To create surface requires doing work, and this is measured by the surface tension of the liquid, or

$$\sigma = \text{surface tension} = \begin{pmatrix} \text{work needed to} \\ \text{create a unit} \\ \text{of fresh surface} \end{pmatrix} \quad [\text{N} \cdot \text{m/m}^2] = [\text{N/m}] \tag{4-5}$$

For   water: $\sigma = 0.072 \text{ N/m}$

   ethanol: $\sigma = 0.024 \text{ N/m}$

   mercury: $\sigma = 0.470 \text{ N/m}$

So to create fresh surface of area A requires doing work, given by

$$\boxed{W = \int_0^A \sigma \, dA \qquad [\text{N/m} \cdot \text{m}^2 = \text{J}]} \tag{4-6}$$

If we were the size of little insects, this type of work would be very important in our lives; for some of us older folks, we'd have to call 911 if a drop of rain trapped us at its surface.

Also note that water in a glass has a single surface or air-water interface; however, a water bubble in air has two surfaces, an inside and an outside surface (Figure 4–5). Consequently, an insect would find it easier to extricate itself from the air-water interface of a glass of water than from an air-water-air interface of a bubble in air.

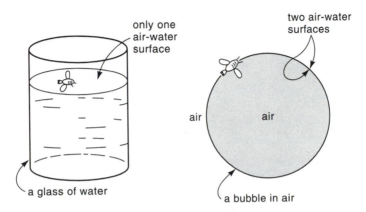

Figure 4–5.

EXAMPLE 4–2.   The uninvited beer drinker

A thirsty fly lands in my glass of warm flat beer, takes a sip, and now tries to leave. How much work does it need to do to extricate itself from the air-beer interface?

Data   The cross-sectional area of the fly in contact with the liquid is 25 mm², the beer is flat with surface tension $\sigma = 0.048$ N/m.

Solution

The work required to recreate the 25 mm² of surface as the fly leaves is

$$W = \int \sigma \, dA = \left(0.048 \frac{N}{m}\right)(25 \text{ mm}^2)\left(\frac{1 \text{ m}^2}{10^6 \text{ mm}^2}\right)\left(\frac{1 \text{ J}}{1 \text{ N·m}}\right)$$

$$= \underline{1.2 \times 10^{-6} \text{ J}} \quad \longleftarrow$$

5. Elastic Work

Consider an elastic springy bar that is pushed from position y_1 to y_2, both deflections small (Figure 4–6).

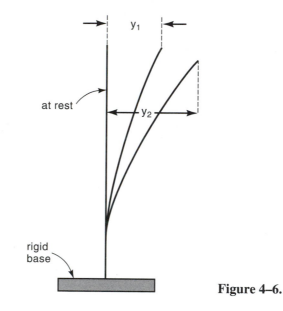

Figure 4–6.

Hooke's law says that the force needed to do this is proportional to the deflection y; thus, the greater the deflection the larger the force needed. In symbols, then,

$$F = K_s y \qquad [J]$$

Hooke's modulus, N/m — *deflection, m* (4–7)

So to deflect the bar from position y_1 to position y_2 requires doing work, given by

$$W = \int_{y_1}^{y_2} F\,dy = \int_{y_1}^{y_2} K_s y\,dy = \frac{K_s}{2}\left(y_2^2 - y_1^2\right) \qquad [J] \tag{4–8}$$

A coiled spring is analyzed this way.

6. Other Forms of Work

There are a number of other forms of work, such as chemical and magnetic. We will introduce these at the proper place in this treatment.

B. HEAT

There is little to be said here, except that heat can be added to the system or object

- by contact with a hotter object which loses energy, or
- by radiation from a hotter object

and can only be removed

- by contact with a cooler object, or
- by radiation to cooler surroundings.

The amount of heat flowing is measured in terms of joules, the rate at which heat is flowing is measured in terms of watts, or joules per second.

PROBLEMS

1. A 9 mm Xian short-barreled experimental pistol is fired, and the explosive gases push the bullet out the barrel of the gun. The pressure is carefully recorded as the bullet accelerates to the exit. Here is a sample of the data collected:

Position of bullet, or distance down the barrel, cm	0	2	3	4	5	6	7	8
Pressure in gun chamber behind the bullet, bar	23	25	24	22	17	10	6	4

bullet at start, just after trigger is pulled

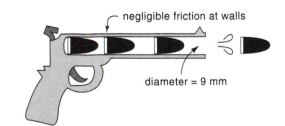

negligible friction at walls

diameter = 9 mm

What is the work done on the bullet?

2. Gas in a cylinder expands by pushing on a piston. At the same time the temperature of the whole setup is changed by adding or removing heat. The data below shows what happens.

Pressure pushing the piston, bar	Volume of gas in the cylinder, lit	Temperature of gas and cylinder, °C
7	10	68
5	12	19
4	14	0
4	16	39
5	18	166
7	20	409

How much work is done by the gas on the piston in this operation?

3. At the 13th Annual Calapooya Flats Soap Bubble Championships held on midsummer day, Jordana blew a giant 83 cm bubble, breaking all previous records. She did this with a specially formulated soap solution having a surface tension of $\sigma = 0.026$ N/m. Find the work received by the soap solution in generating this bubble.

4. A 12-volt battery receives a quick 20-minute charge during which time it receives a steady 50 amps. In this period the battery gets hot and loses 120 kJ of heat. Find the resulting change in internal energy of the battery.

CHAPTER 5

POTENTIAL ENERGY

The potential energy of an object in a field of force, whether gravitational, electric, or magnetic, is the object's extra energy because of where it is in the field. Moving an object from here to there could change its potential energy.

But what is a field? To put it simply, a field of force is a region of space where forces are present. In this chapter we focus on energy changes of a system in a gravitational field. This is represented by the second term in Eq. 3–2.

$$\underset{\text{make} = 0}{\cancel{\Delta U}} + \Delta E_p + \underset{\text{make} = 0}{\cancel{\Delta E_k}} = Q - W \qquad (3\text{–}2)$$

potential energy

The energy equations developed here apply analogously (for the same geometry) to the other kinds of fields, the electric and the magnetic.

A. CONSTANT GRAVITY

By constant gravity we mean that the field forces are constant in the region of space of our interest. This represents moving the system or object somewhere close to the earth's surface, say as high as a jet plane flies, maybe a bit higher (Figure 5–1). In this region the field potential, represented by the value of the acceleration of gravity "g," can be considered to be constant, a situation that we normally meet.[1]

---

[1]We say that the gravitational field, as represented by the value of g, is constant at or near Earth's surface. But this is only an approximation. Also, because Earth is spinning, the centrifugal force lowers g, especially near the equator. In fact, g varies from about 9.78 to about 9.83 m/s² at Earth's surface. The world standard, measured at sea level at 45° north latitude is g = 9.806 65 m/s². But who wants to work in galoshes at sea level? In this book we take g = 9.80 m/s², which means usually a short distance above sea level. Problem 1 calculates what we mean by "short distance."

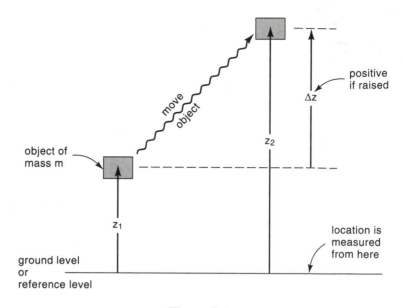

Figure 5–1.

Now, from Newton's law the force of attraction between the earth and any object is

$$F = \frac{mg}{g_c} \quad [N]$$

(5–1)

where *of object* — and — *9.8 m/s² in the vicinity of the earth's surface*

where $g_c = 1$ kg·m/s²·N.[*]

So the first law gives, for negligible velocity effects, and for no change in internal energy, from Eq. 3–2

$$\cancel{\Delta U} + \Delta E_p + \cancel{\Delta E_k} = Q - W$$

where the two cancelled terms $= 0$

which becomes

$$\Delta E_p = Q - W = \int_{z_1}^{z_2} F \, dz = \int_{z_1}^{z_2} \frac{mg}{g_c} \, dz$$

---

[*]In the SI system one can dispense with "g_c" without harm. However, it is included so that engineers and scientists in Burma, the United States, and Ghana who still use the English system of units can use the equations being developed here. For SI users, just ignore g_c, since its value is 1 kg·m/s²·N.

or

$$\Delta E_p = Q - W = \frac{mg}{g_c}(z_2 - z_1) = \frac{mg\,\Delta z}{g_c} \qquad [J] \qquad (5\text{-}2)$$

of system

B. CHANGING GRAVITY

A changing gravity is involved when the force of attraction between the two bodies is not constant, but depends on their changing distance from each other (Figure 5–2). So for any two nonoverlapping bodies this force of attraction is given by Newton's universal law of gravitation

Newton's universal gravitation constant
$6.7 \times 10^{-11}\ N \cdot m^2/kg^2$

$$F = G\,\frac{m_1 m}{r^2} \qquad [N] \qquad (5\text{-}3)$$

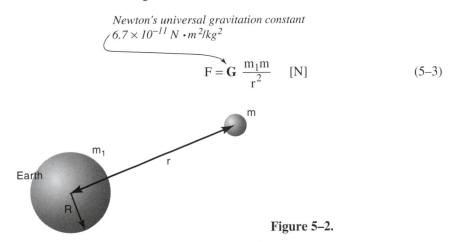

Figure 5–2.

Combining with Eq. 5–1 gives the local value of g for body m when it is a distance r away from m_1

$$g_{local} = \frac{g_c\,G\,m_1}{r^2} \qquad (5\text{-}4)$$

At Earth's surface (Figure 5–3)

$$g_{surf} = \frac{g_c\,G\,m_1}{R^2} \qquad (5\text{-}5)$$

thus

$$\frac{g_{local}}{g_{surf}} = \frac{R^2}{r^2} \qquad (5\text{-}6)$$

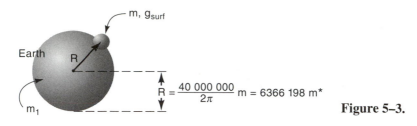

$$R = \frac{40\ 000\ 000}{2\pi}\ m = 6366\ 198\ m*$$

Figure 5–3.

So when a small body m moves from r_1 to r_2 from a second larger body (center to center distance) (Figure 5–4), the first law with Eq. 5–1 tells us that

$$\Delta E_p = Q - W = \int_{r_1}^{r_2} F\ dr = \int_{r_1}^{r_2} \frac{m\ g_{local}}{g_c}\ dr$$

and with Eq. 5–6

$$\Delta E_p = \int_{r_1}^{r_2} \frac{m\ g_{surf}R^2}{r^2}\ dr = \frac{m\ g_{surf}R^2}{g_c}\left(\frac{1}{r_1} - \frac{1}{r_2}\right) \quad [J]$$

*the system gains potential
energy when it moves
away from the earth*

(5–7)

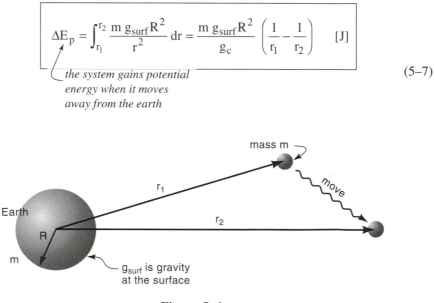

Figure 5–4.

*Incidentally, the standard unit of length, the meter, was chosen by the French Academy about 200 years ago to be 1/10 000 000 the distance from the equator to the pole. To find what that distance actually was, expeditions were sent to northern Norway and also to Ecuador. Their measurements showed that Earth was not a perfect sphere but was somewhat pear shaped, flattened at the South Pole. At its equator its circumference is about 40 074 km with deviations of up to 20 km in actual radius from a sphere. In this book we will chose 40 000 km as Earth's circumference because it is a reasonable working average and is easy to remember.

Let's see how g_{surf} changes with planet size and planet density. For this Eq. 5–8 gives

$$g_{surf} = \frac{g_c \, G \left(\frac{4}{3} \pi R^3 \right) \rho}{R^2} \quad \text{or} \quad g_{surf} \propto R \, \rho \qquad (5\text{--}8)$$

EXAMPLE 5–1.   Garbage disposal

There are discussions about getting rid of very dangerous radioactive waste by ejecting it from Earth and letting it fall into the sun. How much potential energy must be added to this deadly garbage to tear it free from Earth's gravitational field?

Solution

To lift one kilogram of this garbage, the increase in potential energy is given by Eq. 5–7, or

measured at R, at Earth's surface

$$\Delta E_p = \frac{m \, gR^2}{g_c} \left(\frac{1}{r_1} - \frac{1}{r_2} \right)$$

$$= \frac{m \, g \, R}{g_c}$$

$$= \frac{(1 \text{ kg}) (9.8 \text{ m/s}^2) \left(\dfrac{40\,000\,000 \text{ m}}{2\pi} \right)}{(1 \text{ kg·m/s}^2 \text{·N})} = \underline{62.4 \times 10^6 \text{ J}} \quad \longleftarrow \; per \; kg$$

Note.   At 10¢ per kW·hr this would cost (just the potential energy cost)

$$\left(\frac{62.4 \times 10^6 \text{ J}}{\text{kg}} \right) \cdot \left(\frac{\$0.10}{\text{kW·hr}} \right) \left(\frac{\text{hr}}{3600 \text{ s}} \right) \left(\frac{1 \text{ kW}}{1000 \text{ J/s}} \right) = \$1.7/\text{kg}$$

C.   WITHIN A SPHERICAL BODY OF CONSTANT DENSITY

Where the field potential is not described by $F \propto 1/r^2$, you must use the correct expression. For the situation considered here physics tells that the force of attraction between two bodies when one body is within the other is given by

$$\overset{\textit{a constant}}{\underset{\downarrow}{F \ = \ k r}}$$

(5–9)

At the surface of the larger body of radius R and surface gravity g_{surf} we have

$$F_{surf} = kR$$

(5–10)

With Eq. 5–1 we get

$$\frac{F}{F_{surf}} = \frac{r}{R} = \frac{g_{local}}{g_{surf}}$$

(5–11)

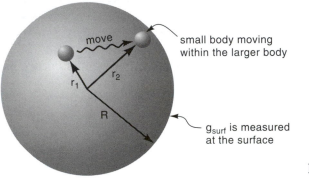

move

small body moving
within the larger body

r_1

r_2

R

g_{surf} is measured
at the surface

Figure 5–5.

So when a body of mass m is moved from r_1 to r_2 in a larger body of radius R (Figure 5–5), the first law says that

$$\Delta E_p = Q - W = \int_{r_1}^{r_2} F \, dr = \int_{r_1}^{r_2} \frac{m \, g_{local}}{g_c} \, dr$$

(5–12)

and with Eq. 5–11

$$\boxed{\Delta E_p = \int_{r_1}^{r_2} \frac{m \, g_{surf} \, r}{g_c R} \, dr = \frac{m \, g_{surf}}{2 \, g_c R} \left(r_2^2 - r_1^2 \right) \quad \text{[J]}}$$

*in moving out from the center of the
large sphere, but still within the sphere,
the system gains potential energy*

(5–13)

Comment.   Sections B and C tell us that as we leave Earth the attractive force (and the local g value) varies as the reciprocal square of the center-to-center distance between object and Earth. On the other hand, as we drill down into the uniform Earth the attractive force decreases linearly to zero at the center of Earth. Thus we have the situation shown in Figure 5–6.

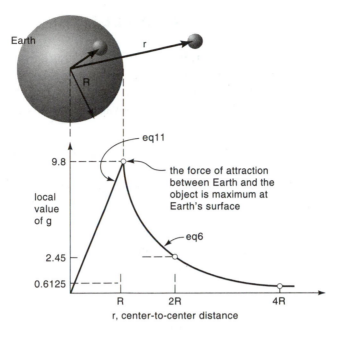

Figure 5–6.

Equation 5–13 only applies to an Earth of constant density. But this does not describe our planet, since recent studies show that the central core has very high density, the crust very low, as shown in Figure 5–7. See C. Tsuboi, *Gravity* (London: Allen and Unwin, 1979).

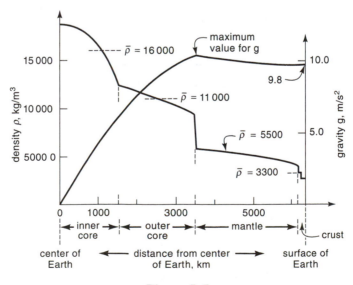

Figure 5–7.

PROBLEMS

1. The value of g varies from place to place on Earth's surface, from 9.78 m/s^2 at the equator to 9.83 at the poles. The standard value adopted by science is set as 9.806 65 m/s^2, the value at about 45° latitude. In this book we take the value of g to be 9.80 m/s^2. What elevation does this correspond to when g = 9.806 65 m/s^2 at sea level?

2. A 1 kg mass is to be brought to the surface from the bottom of a hole drilled to the center of Earth. How much work is needed to do this?
   Data   The circumference of Earth is 40 000 km. Assume a uniform density for Earth.

3. In returning to Earth from exploring the surface of Mars, the space probe must have enough energy to overcome the pull of Mars's gravitational field. How much energy does this amount to (in J/kg)?
   Data   The diameter of Mars is 6787 km and its surface gravity is 38% of Earth's.

4. I am told that NASA is toying with the idea of constructing a permanent moon base halfway between the moon's surface and its center, around year 2055. Diamonds as big as grapefruit are expected to be found at the moon's center since the pressure is very high there. One purpose for this station is to drill to the center of the moon and recover these priceless gems. How long would it take for a 100% efficient 100W motor to bring a 10 kg load from the moon's center to this halfway station?
   Data   Our moon's diameter is 3476 km, its density is approximately constant and equal to Earth's crust, and its surface g value is 16% that of Earth.

5. Find the increase in potential energy when a 1 kg mass is raised 1000 m above Earth's surface where g = 9.80 m/s^2.

   (a) Assume a constant value of g.

   (b) Account for changing g (more accurate solution).

6. Communication satellites are placed in stationary orbit 37 000 km above Earth's surface. This is done in two steps: first, up to 1000 km in a mother rocket, and then the rest of the way in a second-stage rocket. How much potential energy has the satellite gained in the second-stage ascent?

7. Rider Haggard, a wildly popular science fiction writer of the past, wrote a story about explorers who built a "worm" for burrowing into Earth's interior. To their astonishment, they discovered that Earth was hollow. They

had landed in Pellucidar, a foggy land of continents and oceans, with unimagined vegetation, terrifying creatures, and most beautiful damsels—in distress, of course.

The diameter of Pellucidar was exactly one-half that of Earth itself. Rider Haggard automatically assumed that the value of g at Pellucidar was no different from that at Earth's surface. Based on the assumption of a uniform density Earth, I wonder whether this is reasonable? Please determine what he should have assumed.

Hint   A little bit of arithmetic may be helpful, as follows:

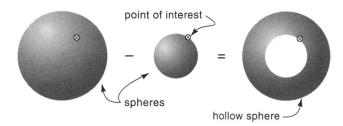

8. Astrologers claim that gravitational forces of the planets can greatly influence our moods and character. For example, lunatic behavior pops out at full moon; Mars was overhead when you were born and that's why you have a tendency to aggression; and so on.

In *The Jupiter Effect* by Gribbin and Plagemann (New York: Random House, 1974) it is suggested that when that giant planet is close to earth, its gravitational effect could lead to all sorts of consequences, trigger earthquakes, affect a newborn's personality, and so on. Were you born at that time? How inauspicious.

In a discussion on astrology on National Public Radio a scientist scoffed at all this nonsense and said that a nurse holding the newborn ($1/2$ m away, center-to-center distance) has a greater gravitational effect on a child than does Jupiter.

Please check and see who is correct; in effect, which has a greater gravitational effect on the newborn, nurse or Jupiter?

Data   Jupiter: mass $= 2 \times 10^{27}$ kg

distance from sun $= 780 \times 10^9$ m

Earth: distance from sun $= 150 \times 10^9$ m

9. Cavendish determined the value of the gravitational constant **G** by measuring the twist of a very fine fiber that held two balls A and B when two other

larger balls C and D are placed close by and then removed, as shown in the sketch below.

Knowing **G** allowed him to determine the mass and density of the earth. Can you repeat Cavendish's calculations, and evaluate the mass and density of the earth given the value of **G** and information reported in this chapter?

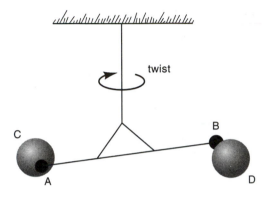

CHAPTER 6
KINETIC ENERGY

The energy of motion of a moving object is represented by its kinetic energy, E_k. Changing the velocity of the object changes its kinetic energy, ΔE_k. In this chapter we consider quantitative measures for ΔE_k in the general energy expression, while keeping E_p and U constant.

$$\textit{make} = 0 \qquad \textit{make} = 0$$
$$\Delta E_p + \Delta E_k + \Delta U = Q - W \tag{3-2}$$

This shows that heat and/or work accompany changes in kinetic energy, and again we will see that there are a number of cases to consider.

A. LINEAR MOTION AT NOT VERY HIGH VELOCITY
(v < 10 000 km/hr)

The kinetic energy of a moving object when referred to the energy of the object at rest is given by

$$E_k = \int_0 F \, dx = \int \frac{m\,a}{g_c} \, dx \tag{6-1}$$

acceleration

This expression has two variables, a and x, and cannot be integrated unless they can be reduced to one variable. We do this most simply by drawing on the findings of Galileo's kinematics, derived before the age of Newton. At constant acceleration Galileo showed that

$$\mathbf{v}^2 = 2\,a\,x$$

and, on differentiating

$$2\,\mathbf{v}\,d\,\mathbf{v} = 2\,a\,dx \tag{6-2}$$

Inserting Eq. 6–2 into Eq. 6–1 gives

$$E_k = \int_0^v \frac{m\,a}{g_c} \frac{2\,v}{2\,a}\,dv = \frac{m\,v^2}{2\,g_c}$$

(6–3)

So when the object's velocity increases from v_1 to v_2, its energy change is

$$\Delta E_k = Q - W = \frac{m}{2\,g_c}\left(v_2^2 - v_1^2\right)$$

(6–4)

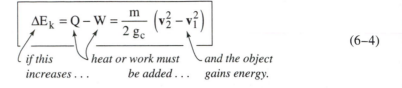

if this   *heat or work must*   *and the object*
increases . . .   *be added . . .*   *gains energy.*

B. ROTATIONAL MOTION AT NOT VERY HIGH VELOCITY

From mechanics the kinetic energy of a rotating body (Figure 6–1) is given by

$$E_k = \frac{1}{2\,g_c}\int v^2 dm = \frac{\omega^2}{2\,g_c}\int r^2 dm$$

(6–5)

$v = \omega r$

*rotational moment
of inertia = I, kg·m²*

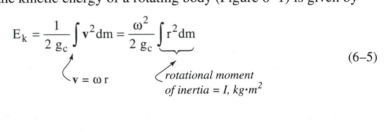

Figure 6–1.

or

$$E_k = \frac{\omega^2 I}{2\,g_c}, \quad \left[\frac{\left(1/s^2\right)\left(kg\cdot m^2\right)}{\left(kg\cdot m/s^2\cdot N\right)} = N\cdot m = J\right]$$

So when the angular velocity changes from ω_1 to ω_2

$$\Delta E_k = Q - W = \frac{\left(\omega_2^2 - \omega_1^2\right) I}{2\,g_c}$$

(6–6)

For an axially spinning object of mass m and radius R (Figure 6–2) the moment of inertia is

$$
\left.
\begin{aligned}
&\text{for a sphere:} && I = \frac{2}{5}\, m\, R^2 \\[2ex]
&\text{for a disc:} && I = \frac{1}{2}\, m\, R^2 \\[2ex]
&\text{for a ring:} && I = m\, R^2 \\[2ex]
&\text{for a spinning} && \\
&\text{clump of matter} && I = m\, R^2 \\
&\text{at distance R:} &&
\end{aligned}
\right\}
\qquad (6\text{–}7)
$$

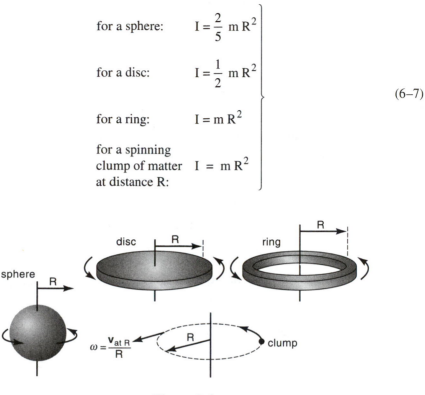

Figure 6–2.

EXAMPLE 6–1. Comparing batteries

Sears' top-of-the-line automobile battery, the Die Hard (72 month warranty) can put out 25 amps at 12 volts for 135 minutes at 27°C. Let us compare this with the energy that can be put out by a flywheel storage system. Here the carbon fiber reinforced disk spins in a vacuum chamber while being levitated in a properly designed magnetic field.

Data   The radius of gyration of the 8 kg disk is 20 cm from its axis, and when fully charged rotates at 60 000 rpm; however, only 85% of the stored kinetic energy can be recovered as electricity.

Solution

Let us calculate the useful stored energy of these devices.

For the lead acid battery, from Eq. 4–3

$$\text{Stored energy} = \mathbf{V} \cdot \mathbf{A} \cdot \text{time}$$

$$= (12 \text{ volts}) (25 \text{ amps}) \left(135 \text{ min} \times \frac{60 \text{ s}}{\text{min}} \right)$$

$$= 2430\,000 \text{ J} = 2.43 \text{ MJ}$$

For the flywheel storage system, from Eq. 6–6

$$\text{Stored energy} = E_k = \frac{\omega^2 I}{2\,g_c} = \left(\frac{\text{radians}}{\text{s}} \right)^2 \left(\frac{m\,R^2}{2\,g_c} \right)$$

$$= \left[2\pi \left(\frac{60\,000 \text{ radians}}{\text{min}} \right) \left(\frac{\text{min}}{60 \text{ s}} \right) \right]^2 \frac{(8 \text{ kg}) (0.2 \text{ m})^2}{2(2 \text{ kg} \cdot \text{m/N} \cdot \text{s}^2)}$$

$$= 3158\,273 \text{ N} \cdot \text{m}$$

and at 85% efficiency

$$\left(\begin{array}{c} \text{Useful stored} \\ \text{energy} \end{array} \right) = (3158\,273 \text{ J}) (0.85) = 2.68 \text{ MJ}$$

The flywheel battery can store

$$\frac{2.68 - 2.43}{2.43} \times 100 = \underline{\underline{10.3\% \text{ more energy}}} \quad \longleftarrow$$

For more on this subject read R. F. Post and S. F. Post, Flywheels, *Scientific American*, 229, 17 (1973).

C.   AT VERY HIGH VELOCITY, APPROACHING THE SPEED OF LIGHT

Einstein showed that the total energy of a moving object, with $\Delta E_p = 0$, is given by

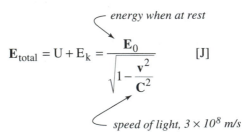

$$\mathbf{E}_{\text{total}} = U + E_k = \frac{\mathbf{E}_0}{\sqrt{1 - \dfrac{\mathbf{v}^2}{\mathbf{C}^2}}} \qquad [\text{J}]$$

energy when at rest

speed of light, 3×10^8 m/s

where the energy of the object at rest is

$$E_0 = \frac{m\,C^2}{g_c} \qquad (6\text{--}8)$$

and

$$C = 300\,000 \text{ km/s}, \quad \text{speed of light in a vacuum.}$$

So the kinetic energy of the fast moving object is

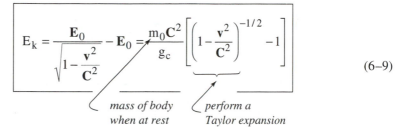

$$E_k = \frac{E_0}{\sqrt{1 - \dfrac{v^2}{C^2}}} - E_0 = \frac{m_0 C^2}{g_c}\left[\left(1 - \frac{v^2}{C^2}\right)^{-1/2} - 1\right] \qquad (6\text{--}9)$$

mass of body   perform a
when at rest   Taylor expansion

Now the Taylor expansion on the square root term gives

$$(1 + x)^n = 1 + nx + \frac{n(n-1)}{2!}\,x^2 + \frac{n(n-1)\,(n-2)}{3!}\,x^3 + \dots$$

and applying it to Eq. 6–9 gives

$$(1 - a)^{-1/2} = 1 + \frac{1}{2}a + \frac{3}{8}a^2 + \frac{5}{16}a^3 + \dots$$

or

$$E_k = \frac{m_0 C^2}{g_c}\left[\left(1 + \frac{1}{2}\frac{v^2}{C^2} + \frac{3}{8}\left(\frac{v^4}{C^4}\right) + \frac{5}{16}\left(\frac{v^6}{C^6}\right) + \dots\right) - 1\right]$$

or

$$E_k = \frac{m_0}{g_c}\left[\frac{1}{2}v^2 + \frac{3}{8}\frac{v^4}{C^2} + \frac{5}{16}\frac{v^6}{C^4} + \dots\right] \qquad (6\text{--}10)$$

this is the ordinary   even at Mach 10, or 10 000 km/hr
low velocity   this second and subsequent terms
expression, eq 4   are negligible

What this shows is that our ordinary kinetic energy expression, Eq. 6–3, is just the low velocity approximation to the more general Eq. 6–9.

Nuclear reactions spit out enormous amounts of energy. This occurs by having elementary particles—neutrons, protons, electrons, and so on—shoot out of the nucleus of atoms at unbelievable velocities, close to the speed of light. They have enormous amounts of kinetic energy; in slowing down, their kinetic energy turns into internal energy and heat. Thus, the enormous energy release of nuclear reactions and atomic bombs really comes from reducing the E_k of the particles from their high initial velocity. For example, if one gram of material, moving at 0.99 **C** is stopped, it will release as much energy as 17 000 m^3 of gasoline. Imagine, slowing down four ordinary sized postage stamps (1 gm) would allow you to drive your car 3500 times around the world, over land and water.

PROBLEMS

1. Record books claim that Bob Feller had the fastest fast ball in professional baseball. Well, I've got a pretty fast left arm myself. Why, I've thrown a baseball 40 m into the air. How fast is my fast ball, in m/s and in miles/hr?
   Note: Please ignore air friction. It should not have much effect at these velocities.

2. The *Corvallis Gazette-Times*, Monday, December 1, 1986, reported that the giant experimental twin-bladed wind turbines, the largest rotating machines in the world (diameter of the rotor = 91 m), built by Boeing Company and located at Goldendale, Washington, were being scrapped. Do you need one in your backyard? They're a bargain. The news article also reports that the average power output of each of these monster machines is 2.5 MW. How does this compare with the best possible?
   Data   The theoretical maximum utilization of the kinetic energy of wind that approaches the turbine blades is 59%. The average air temperature is 10°C and the average wind velocity at Goldendale is 24 km/hr.

3. Soapbox derby racing is a fiercely competitive sport, with national competitions and great prizes and honors awarded for success. The way it is run, the four-wheeled vehicles start at rest at the top of a hill and coast down a smooth straight track. The person who arrives first at the bottom is the winner. We plan an entry that runs on bicycle wheels. Should we use large diameter or small diameter wheels of identical mass?
   Data   Tires are pumped hard, so our vehicle can be considered frictionless.

4. Sandia National Laboratories, New Mexico, created an experimental gun that can blast a projectile at the highest velocity ever reached on earth by any object larger than a speck of dust.

In a vacuum, the 20 m long gun, the Enhanced Hyper Velocity Launcher (EHVL), is reported as able to accelerate a small object to 16 km/s. For a 1 gm bullet, weighing about the same as four regular postage stamps,

(a) find the energy imparted, and

(b) find the theoretical power required to reach this velocity.

CHAPTER 7

INTERNAL ENERGY U AND ENTHALPY H

A. INTERNAL ENERGY

As shown in Figures 3–1, 3–2, and 3–3 and in Eq. 3–2, adding heat or work to a system can change its kinetic, potential, or internal energy. Let us consider this internal energy change.

What is internal energy? It is the energy of the molecules themselves, the attractive forces that keep the molecules of liquids and solids from flying apart, the kinetic energy of fast moving gas molecules as they buzz about, the extra energy required to create a surface, and so on. Although it is rather awkward to say exactly what internal energy is, we can still measure changes in internal energy without difficulty.

Consider an analogy by asking how much money a bank has. This is difficult to evaluate; however, finding the difference between what it has today and what it had yesterday is much easier to determine. Just draw a dotted line around the bank and account for what crosses this boundary, as shown in Figure 7–1.

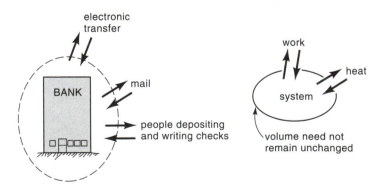

Figure 7–1.

We do the same thing here with internal energy. We draw an imaginary boundary around the system and we note the heat and work that crosses this boundary. We also account for the changes in potential and kinetic energy of the system. What is then left unaccounted is the change in internal energy. Thus, from Eq. 3–2

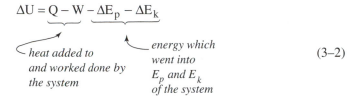

$$\Delta U = \underbrace{Q - W}_{} \; \underbrace{- \Delta E_p - \Delta E_k}_{}$$

heat added to
and worked done by
the system

energy which
went into
E_p *and* E_k
of the system

(3–2)

B. ENTHALPY, A USEFUL MEASURE

A weight scale in a shopping mall doesn't just register the weight of a person. It always weighs the person plus clothes. Similarly, when you want to evaluate the energy of an object of volume V you have to remember that the object had to push the surroundings out of the way to make room for itself. With pressure p on the object, the work required to make a place for itself is

$$W = \int_0^V p \, dV = pV$$

This is so with any object or system, and this work may not be negligible. Remember that at one bar the force of the atmosphere on each square meter is equivalent to that exerted by a mass of about ten tons, and this is not to be sneezed at.

Thus, the total energy of a body (consider a gas in a container) is its internal energy plus the extra energy it is credited with by having a volume V at pressure p. We call this total energy the enthalpy H. For a system of volume V and having a pressure p we define the enthalpy as

$$H = U + p \, V$$

internal ↗ ↖ *pressure*
energy *in system*

(7–1)

and between states 1 and 2 the difference in enthalpy is

$$\Delta H = H_2 - H_1 = \Delta U + \Delta(p \, V)$$
$$= (U_2 + p_2 \, V_2) - (U_1 + p_1 \, V_1)$$

(7–2)

The pV term represents a sort of potential energy.

Warning. We certainly will meet situations where the pressure in the system may not be that of the surroundings. In that case remember that you should not use the surrounding pressure to evaluate the enthalpy. Why? Because we did not define the enthalpy that way.

C. MORE ON THE FIRST LAW EQUATION FOR BATCH SYSTEM

When a system moves from one state to another, it may produce work to turn a shaft, generate electricity, and so on. This is called shaft work, W_{sh}. The system may also expand, as when liquid water turns to steam. This is called pV work, W_{pV}. Thus the total work produced is of two types

$$W = W_{sh} + W_{pV} \qquad (7\text{–}3)$$

and so in general Eq. 3–2 for closed systems can be written as

$$\boxed{\Delta U + \Delta E_p + \Delta E_k = Q - W_{sh} - W_{pV} \quad [J]} \qquad (7\text{–}4)$$

any work that isn't pV work, $\quad\quad -\left(\dfrac{}{}\int p\,dV\right.$
thus shaft, electrical, etc.

There are two useful special cases for this general expression

- At constant volume

$$\Delta U + \Delta E_p + \Delta E_k = Q - W_{sh} \qquad (7\text{–}5)$$

- At constant pressure

$$\Delta U + \Delta E_p + \Delta E_k = Q - W_{sh} - \int \overset{constant}{p\,dv}$$

$$= Q - W_{sh} - p_2 V_2 + p_1 V_1$$

and with Eq. 7–2

$$\Delta H + \Delta E_p + \Delta E_k = Q - W_{sh} \qquad (7\text{–}6)$$

D. SHOULD WE USE ΔU OR ΔH?

For constant volume batch systems we usually find it more convenient to use ΔU equations. However, you will see that for constant pressure batch systems and for all types of flow systems, the ΔH equations are always preferred.

Since equations written in terms of enthalpy are usually more useful for engineering purposes than the corresponding ones written in terms of internal energy, many textbooks don't even bother to include internal energy values in their property tables. If needed, just note that u for a system can be gotten from h, p, and v according to

$$u = h - pv \cdots \text{[J/mol or J/kg]} \quad \text{or} \quad U = H + pV \cdots \text{[J]}$$

You may be uncomfortable at this point with the concept of enthalpy. However, I assure you that you will soon be as much at ease with it as you would be with using chopsticks after a few tries.

E. STANDARD STATES FOR U AND H

Now the absolute value of the internal energy of any material at rest, indicated by subscript zero, is given by Einstein's famous expression[1]

$$U_0 = \mathbf{E}_0 = \frac{m_0 \mathbf{C}^2}{g_c} \qquad \text{(6–8) or (7–7)}$$

speed of light = 3 × 10⁸ m/s

Hence, the change in internal energy of a material at rest is simply accompanied by a corresponding change in mass Δm of the object

$$\Delta U_0 = \Delta \mathbf{E}_0 = \frac{\Delta m_0 \mathbf{C}^2}{g_c} \qquad \text{(7–8)}$$

This expression says that if you heat an object it gets heavier; cool it and it gets lighter.

Unfortunately, this expression is not practical for our everyday use. Let us explain why. If one ton of water is heated from 10°C to 90°C the increase in energy, in terms of joules, would be enormous. However, the increase in mass would be so small as to be unmeasurable even by our most sensitive balances. So to use Eq. 7–8 would be impractical.

Instead, for practical purposes, we arbitrarily pick some convenient standard state, we call the internal energy or enthalpy zero at that state, and go from there with positive and negative values relative to that state. We pick

---

[1]Einstein came up with this seemingly magical expression in 1905 while developing his special theory of relativity. The Nobel prize was awarded to him not for this theory but for something else he did that very same year. Yes, 1905 was quite a year for this 25-year-old genius. Actually, he published the theory in 1905, but he must have been quite a bit younger (maybe about the age of most of you readers) when he first dreamed about these wild ideas.

For water (liquid) $U = 0$ at $T = 0.01°C$, $p = 611.3$ Pa, the triple point[2]

For ammonia (liquid) $H = 0$ at $T = -40°C$, $p = 71\ 770$ Pa

For refrigeration we have a slight problem. As mentioned elsewhere in this book, our long accepted working fluid, Freon-12, has recently been banned and is being replaced by chlorine-free HFC-134a for use in home refrigerators, auto air conditioners, and other devices. However, the British and Americans don't seem to speak to each other because they call this material by different names and haven't agreed on the same standard state. ICI calls it Klea-134a, while du Pont calls it Suva-134a. For the standard state of HFC-134a

ICI chooses $h = 100$ kJ/kg
du Pont chooses $h = 200$ kJ/kg $\Big\}$ at $T = 0°C$, $p = 292\ 930$ Pa

The table at the back of this book uses the du Pont values. However, since we always want ΔH, use any table you wish. Just don't take one value for ΔH from the du Pont table, the other value from the ICI table. That would be silly.

PROBLEM

1. An incompressible block of metal is resting on the bottom of my bathtub which is filled with water. I pull the plug and drain the water from the tub. Everything stays isothermal. What happens to the energy and the enthalpy of the block?.

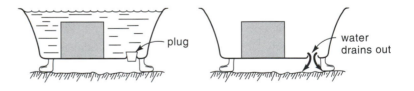

---

[2]The triple point is the unique pressure and temperature at which the solid, liquid, and gas phases of a pure substance all exist in equilibrium with each other. We talk about this in Chapter 12.

CHAPTER 8

ΔU AND ΔH
FOR PHYSICAL CHANGES

By adding heat, work, E_p, and/or E_k to a system, the change in U or H can do a number of things. It can

1. change the temperature of the system (raise the temperature).
2. change the phase of the system (solid → liquid → gas).
3. change the chemical structure ($H_2O → H_2 + 1/2 O_2$)
4. change in atomic structure.

We consider changes 1 and 2 in this chapter and changes 3 and 4 in the next chapter. Note also that u and h refer to a single mole or kg of material while U and H refer to the energy of the whole material.

A. ΔU AND ΔH FOR A TEMPERATURE CHANGE

For a given material, when T rises, U and H increase because $U = f_1(T)$ and $H = f_2(T)$, as shown in Figure 8–1.

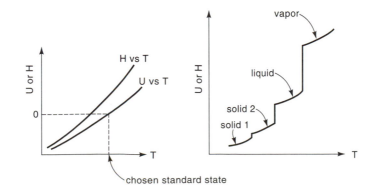

Figure 8–1.

We can write for various changes

$$\text{at constant V:} \quad \Delta U = n \int c_v dt \quad \text{where } c_v = a + bT + \ldots \quad (8\text{--}1)$$

$$\text{at constant p:} \quad \Delta H = n \int c_p dt \quad \text{where } c_p = a' + b'T + \ldots \quad (8\text{--}2)$$

c_v and c_p, in J/mol·K or J/kg·K, are called the *specific heat*, or *specific heat capacity* of the substance, the subscripts letting you know that they stand for constant volume or constant pressure changes.

Remember, Δh and Δu differ in that Δh includes the work needed to push back the atmosphere as the material heats up and expands. For liquids and solids the volume change is very small, hence $\Delta h \cong \Delta u$ and $c_p \cong c_v$.

To illustrate, let us raise the temperature of 1 kg of a gas, a liquid, and a solid from 20 to 30°C and see how important is the $\Delta(pV)$ contribution to the total energy change.

At 1 atm	$\Delta(\mathbf{pv}) = \Delta\mathbf{h} - \Delta\mathbf{u}$	$\Delta\mathbf{u}$	$\Delta\mathbf{h}$
for air, from 20 to 30°C	2850 J/kg	7170 J/kg	10 020 J/kg
for water, from 20 to 30°C	0.1 J/kg	41 840 J/kg	41 840 J/kg
for iron, from 20 to 30°C	0.004 J/kg	4494 J/kg	4494 J/kg

notice how negligible is the pV terms for solids and liquids

For a gas, the work required to push back the atmosphere may not be negligible, and should not be ignored. Think of the airplane accident reported a few years ago when a window on a commercial jet blew out, the outrushing air trapping a nearby passenger, sucking and squeezing him through the hole like toothpaste out a toothpaste tube. Ugh!

B. ΔU AND ΔH FOR A PHASE CHANGE

The energy needed to boil water and form steam has to do two things. First, it has to make the molecules so active and energetic that they fly apart from each other and overcome the attractive forces that had held them close to each other as liquid. Secondly, energy has to be added to push the atmosphere back and make room for the steam. The total energy needed for such a change is called the *latent heat of vaporization*, $h_{\ell g}$ (J/kg or J/mol).

As an example, for 1 kg of liquid water turning into steam at 100°C and 1 atm, the tables at the back of this book gives

The energy to tear the molecules apart:

$$\Delta u_{\ell g} = 2087\ 600 \text{ J/kg}$$

The work required to push the atmosphere back:

$$p\,(v_g - v_\ell) = (101\ 325)\,(1.673 - 0.001\ 044) = 169\ 400 \text{ J/kg}$$

So the latent heat of vaporization is

$$\Delta h_{\ell g} = 2087\ 600 + 169\ 400 = 2257\ 000 \text{ J/kg}$$

Note that one volume of liquid water gives about 1600 volumes of steam, and the energy needed to do this is about 8% of the total. Since one cannot go from liquid to vapor without a volume change, $\Delta u_{\ell g}$ is not a useful quantity and one always uses $\Delta h_{\ell g}$.

For a solid to liquid latent heat, called *heat of fusion*, $h_{s\ell}$ or $H_{s\ell}$, the volume change is usually very small so $\Delta u_{s\ell} \cong \Delta h_{s\ell}$.

The latent heat of vaporization, $\Delta h_{\ell g}$, also changes with temperature, being higher at low temperature. This makes sense because when the molecules of liquid are cold it takes more energy to tear them away from their neighbors and have them fly away as vapor. On the other hand, when the liquid is very hot the molecules are very energetic and buzzing around very actively. So it takes very little extra energy to kick them into the vapor. So for water, as an example

$$\Delta h_{\ell g} = 2454.1 \text{ kJ/kg} = 44.17 \text{ kJ/mol} \quad \text{at } 20°C$$

$$\Delta h_{\ell g} = 893.4 \text{ kJ/kg} = 16.08 \text{ kJ/mol} \quad \text{at } 350°C$$

TABLE 8–1. Useful Simple Approximate Values of c_p, c_v, $\Delta h_{\ell g}$ and $\Delta h_{s\ell}$

For solids and liquids $c_p \cong c_v$,   and for metals $c_p \cong 400$ J/kg·K

For gases $c_p > c_v$,   and for air $c_p \cong 1000$ J/kg·K

Ideal gases

Monomolecular: $c_p = (^5/_2)R = 20.79$ J/mol·k $c_v = (^3/_2)R = 12.47$ J/mol·K

Bimolecular: $c_p = (^7/_2)R = 29.10$ $c_v = (^5/_2)R = 20.79$

Trimolecular: $c_p \cong 4R \cong 33.0$ $c_v \cong 3R \cong 25.0$

For all ideal gases: $c_p = c_v + R$, where $R = 8.314$ J/mol·K

For Water

Ice:	$c_p = 2000$ J/kg·K $= 36$ J/mol·K		
Liquid water:	$c_p = 4814$	$= 75$	$\Delta h_{s\ell,0°C} = 333$ kJ/kg $= 6.0$ kJ/mol
Steam:	$c_p = 1500$	$= 27$	$\Delta h_{\ell g,100°C} = 2250$ $= 40.6$

Note: For more accurate values see the appendix.

EXAMPLE 8–1.   Preparing a cup of tea

I pour 1 liter of water at 20°C into a thermally insulated electrically heated teapot (Figure 8–2), which I then plug in. How long will I have to wait for the water to boil and for the pot to whistle?

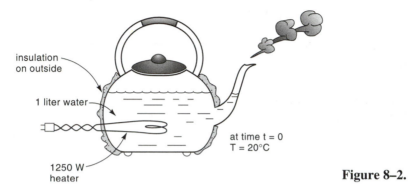

insulation
on outside

1 liter water

1250 W
heater

at time t = 0
T = 20°C

Figure 8–2.

Data   The metal of the teapot is equivalent to 200 cm³ of water, and the label on the teapot tells that its heater is rated at 1250 W.

Solution

Let the kettle and its contents be the system. Then, since the pressure in the system is constant and equal to the atmosphere that is being pushed back as water boils, the first law expression of Eq. 7–6 should be used. This is written and simplified as follows

$$\Delta H + \overset{=0}{\cancel{\Delta E_p}} + \overset{=0}{\cancel{\Delta E_k}} = Q - \overset{=0}{\cancel{W_{sh}}}$$

and with the values in Table 8–1, the amount of heat needed is

$$Q = \Delta H = m\, c_p \Delta T$$

$$= (1.2 \text{ lit}) \left(\frac{1 \text{ kg}}{1 \text{ lit}} \right) \left(4184 \frac{J}{kg \cdot °C} \right) (100°C - 20°C)$$

$$= 401\ 664 \text{ J}$$

With a heating rate of 1250 W

$$\left(\begin{array}{c} \text{The time needed to} \\ \text{boil the water} \end{array} \right) = \frac{401\ 664 \text{ J}}{1250 \text{ J/s}} = 321 \text{ s}$$

$$= \underline{\underline{5 \text{ min } 21 \text{ s}}} \longleftarrow$$

EXAMPLE 8–2.   The maddening whistle

How long will the shrill whistling of the teapot of Example 8–1 last?

Solution

Assume that you don't pull the electric plug and that the whistling only stops when all the water boils away. Then the energy input needed to boil 1 liter of water is, from Table 8–1,

$$\cancel{Q} - W_{sh} = m\,\Delta h_{\ell g}$$

$$= (1\text{ lit})\left(1\frac{kg}{lit}\right)\left(2255\frac{kJ}{kg}\right) = 2255\text{ kJ}$$

approximations

The electric power input is

$$\cancel{Q} - W_{sh} = 1220\text{ W} = 1220\,\frac{J}{s}$$

So the time needed to boil the water

$$t = \frac{\text{energy needed}}{\text{energy input rate}}$$

$$= \frac{2255\,000\text{ J}}{1220\text{ J/s}} = 1841\text{ s}$$

$$= \underline{30\text{ min }41\text{ s}} \quad\longleftarrow$$

EXAMPLE 8–3.   The thermodynamics of a spectacular crash (Figure 8–3)

Here you are tooling along at 200 km/hr in your avocado and puce colored 1923 Hupmobile sedan de luxe when you spy a 1996 Thunderbird 32 cylinder super coming right at you (in your lane, of course) at 200 km/hr. Before you can tip your hat—CRASH. Lots of things happen, one of which is a temperature rise of this instantly created pile of junk. What is this temperature rise, and what are ΔU and ΔH for this catastrophic event?

Data   Each car weighs 2 tons and has an average c_p of 0.5 kJ/kg·K.

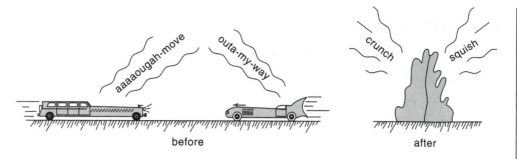

before   after

Figure 8–3.

Solution

Here the masses are equal, so the same thing happens to each car, it comes to rest. So the cars' kinetic energy is transformed into internal energy, or

$$\Delta U + \underset{increase}{\Delta E_p} + \underset{decrease}{\Delta E_k} = \overset{=0}{\cancel{Q}} - \overset{=0}{\cancel{W}} \qquad \text{(3–2) or (7–4)}$$

or

$$m\,c_v(T_2 - T_1) = -\frac{m\left(\overset{=0}{\cancel{v_2^2}} - v_1^2\right)}{2\,g_c}$$

Since the cars are made of solids and liquids $c_v \cong c_p$, we have for each kilogram of cars

$$\left(500\,\frac{J}{kg{\cdot}K}\right)(\Delta T,\ K) = \left[\left(200\,\frac{km}{hr}\right)\left(\frac{1000\ m}{km}\right)\left(\frac{hr}{3600\ s}\right)\right]^2 \Big/ 2\left(1\,\frac{kg{\cdot}m}{s^2{\cdot}N}\right)$$

or

$$500\,\Delta T = 1543$$

$$\text{or}\ \ \Delta T = \frac{1543}{500} = \underline{\underline{3.09°C}} \quad \longleftarrow$$

As for the enthalpy change and internal energy change per kilogram of cars

$$\Delta h \cong \Delta u = c_v \Delta T = \left(500\,\frac{J}{kg{\cdot}K}\right)(3.09\ K) = 1543\,\frac{J}{kg}$$

So for this event that involved two cars, m = 4000 kg, and in total

$$\Delta H \cong \Delta U = m\, c_v \Delta T = (4000 \text{ kg}) \left(1543 \frac{J}{kg} \right) = 6172\,000$$

$$= 6172\,000 \ \frac{J}{2 \text{ cars}}$$

$$= 6.172 \ \frac{MJ}{\text{event}} \quad \longleftarrow$$

PROBLEMS

1. I plan to heat 1 ton (metric) of liquid water at 25°C to its boiling point (100°C) and then heat the vapor to 500°C, all at 1 atm. How many joules of heat are needed to do this?

  (a) Solve using only the approximate values given in Table 8–1.

  (b) Solve using the values in the tables at the back of the book.

2. Calculate the heat required to raise 1 kg of graphite from 0°C to 800°C in an oxygen free environment.

Data	Temperature (K)	c_p (J/kg·K)
	200	420
	400	1070
	600	1370
	800	1620
	1000	1820

3. A 1500 W heater is turned on for 1 ks so as to heat 5 kg of ice at 0°C. If no heat is lost to the surroundings, find the final state of this ice. Is it vapor, liquid, or solid, or is it a mixture of s-ℓ or ℓ-g? Also find the final temperature of this 5 kg of H_2O.

4. A 2 kg metal bar (c_p = 800 J/kg·K), initially at 200°C, is dropped into a tank of water at 27°C. After some time the temperature of both the metal bar and the water are found to be 75°C.

(a) How much has the enthalpy of the metal bar changed during this process?

(b) If all of the energy from the metal bar is transferred to the water, how much water is in the tank?

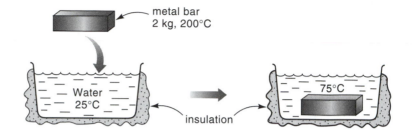

5. *"Cleanliness is, indeed, next to godliness,"* said John Wesley (1703–1791). Well, three people live in our home, and each of us takes an average of two baths or showers each day. Each cleansing uses 32 gal of warm water ($\frac{1}{3}$ cold at 5°C and $\frac{2}{3}$ hot at 72.5°C). An electric hot water heater gives us our hot water, and electricity costs 8.0¢/kW·hr. How much does it cost us per month to rise to this divine state?

6. *Phase change of a salt.* How to store unneeded heat that is produced in low-use periods by electricity generating plants is a problem studied at the University of Illinois (*Chemical Engineering News*, September 1975). The concept involves storing the heat by melting a salt contained in storage structures 100 m square and 15 m deep. Each such structure would hold 220 million kg of sodium nitrate.

   During low energy demand periods pipes inside the tank would carry steam to melt the salt. During peak demand periods the pipes would carry water that would boil as the salt resolidifies.

   A typical large electric power plant has an output of about 1 GW. For what length of time could this salt storage tank absorb and store the whole output of such a power plant? While absorbing and giving up heat the salt stays at its melting point.

   Data   At 1 bar sodium nitrate, $\overline{mw} = 0.085$ kg/mol, melts at 310°C with a latent heat of fusion $\Delta h_{s\ell} = 15.9$ kJ/mol.

7. A mysterious substance X is to be heated from 20°C where it is a solid to 370°C where it is a vapor. What is the enthalpy change in J/kg for this operation?

Data For substance X $\overline{mw} = 0.1$ kg/mol

Melting point = 120°C   Latent heat $\Delta h_{s\ell} = 200\ 000$ J/kg

Boiling point = 170°C   Latent heat $\Delta h_{\ell g} = 30\ 000$ J/mol

$c_{p,s} = 1000$ J/kg·K $c_{p,\ell} = 2000$ J/kg·K $c_{p,g} = 150$ J/mol·K

8. Angel Falls in Venezuela is the world's highest waterfall with a drop of 975 meters. (Niagara is peanuts by comparison—just 50 m.)

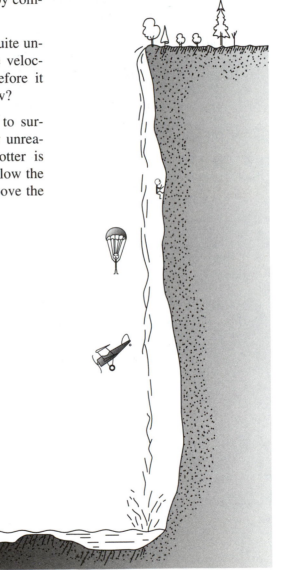

(a) Assuming no friction (quite unreasonable), what is the velocity of the water just before it hits the river down below?

(b) Assuming no heat loss to surroundings (outrageously unreasonable), how much hotter is the water in the river below the falls than in the river above the falls?

9. Captain Schultz,[1] the star high diver of the circus, insists that his bath always be at exactly 37°C, body temperature, and measures the temperature of the bath with a precision thermometer before relaxing in it at the end of his day's work. One terrible evening not so long ago the captain found his bath to be only 36.8°C. His noble countenance glowed with fury as he ordered his trembling valet to move his portable 80 kg tub from his private dressing tent to the foot of the high-dive ladder. For a moment he paused, deep in thought, and then, ignoring the murmurs of the gathering crowd, he mounted the vertical ladder, counting the rungs as he climbed. He stopped at what was obviously a critical height. Disdainfully he let his dressing gown fall to the spellbound throng far below, and launched himself into the air, landing in his tub with such infinite precision that nary a drop of water was lost from the tiny splash. A smile lighted his features as he relaxed in his 37°C tub, while his valet quickly wheeled him back to his tent. Captain Schultz is 170 cm tall, has blond hair, blue eyes, weighs 70 kg. How high did he climb?

80 kg water at 36.8°C

10. P. W. Pipe Co. is Oregon's largest manufacturer of PVC pipe, and at their Eugene plant they make pipes of all sizes up to 60 cm ID and with a wall thickness of 2.5 cm.

   The first step in making this pipe is to mix together and gently cook PVC powder (95%) with a variety of additives—chalk, coloring material, wax as lubricant, and so on—in giant blenders. This product then goes to powerful extruders that squish out the pipe like toothpaste.

   The mixing and heating is done by simply vigorously stirring the powder with giant paddles. What average rate of stirring energy input is needed to heat the half-ton batch of powder from a room temperature of 20°C to 116°C in 5 minutes? Assume no heat loss to the vessel walls.

Data   For the powder being mixed: $c_p = 625$ J/kg·K.

---

[1] Captain Schultz first performed this spectacular high dive on pg 102 of Sussman's *Elementary General Thermodynamics* (Reading, MA: Addison Wesley, 1972).

11. When a high-velocity armor-piercing shell hits its target, such as the armour plating of a tank, its kinetic energy is transformed into internal energy that melts the front part of the shell and the plating in contact with it, thereby allowing it to slide through the melted hole and explode inside the tank.

To successfully pierce, or more correctly, to melt its way through the armor plating, we calculate that the shell has to melt its front half plus double the shell's mass of armor plating. What should be the velocity of the shell to be able to do this?

Data   Take the properties of shell and plating to be that of steel, or

$$c_p = 500 \text{ J/kg} \cdot {}^\circ\text{C}, \ \Delta h_{s\ell} = 15\,000 \text{ J/mol}, \ T_{s\ell} = 1375{}^\circ\text{C},$$

$$\overline{mw} = 0.055 \text{ kg/mol}$$

Assume that anything that doesn't melt doesn't even heat up.

12. Nepal hopes to proudly join the space nations with its "Annapurna" space-craft, which is to carry two Sadhus (holy men) into a 100 km high orbit circling the earth at 5 km/s. This is the same orbit as the U.S. space shuttle.

You are hired to design the heat shield to dissipate 99% of the heat generated by friction as this holy craft returns to earth. However, you must first determine whether the temperature rise in the craft would perchance discomfort the Sadhus. If so, they may need to take a cooler of ice-cold Pepsi with them.

If Annapurna is to come down in a gentle landing at Swayambu, how hot will the craft get?

Data   Assume that $c_p = 500$ J/kg·K for the materials making up the space-craft.

Take a mean value for the acceleration of gravity, $\overline{g} = 9.70 \text{ m/s}^2$.

13. In Rangoon, Burma, I witnessed a primitive but clever way to butt-weld (end-to-end) two metal rods (see sketch). Rod A is attached to a 160 kg disk 1 m in diameter that is made to spin by a gear connection to a bicycle ridden by a young man. This spinning rod is pressed against a stationary rod B, frictional heat causes thin slices of steel to melt as the disk screeches to a halt. How fast (in r.p.m.) must the disk rotate to cause the two 1 cm diameter rods A and B to weld together?

Data   For Burmese steel

Density: $\rho = 7820 \text{ kg/m}^3$

Melting point: $T_m = 1460{}^\circ\text{C}$

Specific heat: $c_p = 460$ J/kg·K

Latent heat of fusion: $\Delta h_{s\ell} = 270\ 000$ J/kg

Layer of steel in each rod that is heated to its melting point = 1.5 mm

Layer of steel in each rod that melts = 10 μm

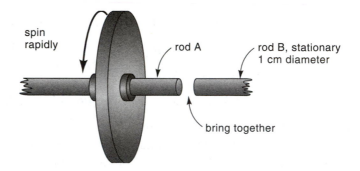

14. The 300 Weatherby Magnum high velocity rifle using 110 gm Spire point bullets shoots these bullets at muzzle velocities as high as 1181 m/s. If shot normal to a silicon nitride wall, if all the kinetic energy of a bullet is completely transformed into internal energy of the bullet, if the bullet is made of lead, what will be the temperature and state of the bullet?

Data For the bullet: $T_{firing} = 20°C$, $c_{p,s} = 135$ J/kg·K

$c_{p,\ell} = 140$ J/kg·K

To melt: $T_{s\ell} = 327°C$ $\Delta h_{s\ell} = 24\ 740$ J/kg

To vaporize: $T_{\ell g} = 1700°C$ $\Delta h_{\ell g} = 850\ 140$ J/kg

CHAPTER **9**

ΔU AND ΔH FOR CHEMICAL AND NUCLEAR REACTING SYSTEMS

Start with a system at temperature T that contains, among other things some A and some B. Then, as an example, shown in Figure 9–1 suppose that n moles of A react with 3n moles of B to produce 2n moles of R, after which the system is returned to temperature T. In general the volume may not be the same before and after reaction.

A.   REACTION IN A CONSTANT VOLUME SYSTEM[1]

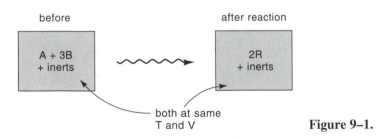

Figure 9–1.

Here we write the first law as

$$Q - W_{sh} = \Delta U_r + \overset{=0}{\cancel{\Delta E_p}} + \overset{=0}{\cancel{\Delta E_k}}$$

(7–5) or (9–1)

we write the subscript r to remind ourselves that the change in U is due to reaction, not temperature or phase change alone.

The term ΔU_r is called the *heat of reaction at constant volume*. More compactly, we write

$$A + 3B \rightarrow 2R \qquad \dots \Delta U_r \; [\text{J/mol A, or J/3mol B, or J/2mol R}]$$

---

[1]Actually the volume and temperature of the system need not be constant throughout the process. It only has to *start* and *end* at the same values, at the time we make an accounting of the energy changes involved.

B.   REACTION IN A CONSTANT PRESSURE SYSTEM (VOLUME MAY CHANGE) (FIGURE 9–2)

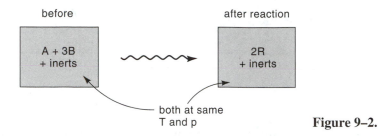

before   after reaction

A + 3B + inerts   2R + inerts

——— both at same
       T and p   **Figure 9–2.**

Here, we write the first law, from Eq. 7–6, as

$$Q - W_{sh} = \boxed{\Delta H_r} + \underset{=0}{\Delta E_p} + \underset{=0}{\Delta E_k} \tag{9–2}$$

called the heat of reaction at constant pressure

More compactly we represent this reaction as follows

$$A + 3B \rightarrow 2R \ldots \Delta H_r \text{ (J/mol A, or J/3mol B, or J/2mol R)}$$

If $\Delta H_r > 0$, it means that H_{after} has to be larger than H_{before}, hence we must *add* energy to the system otherwise the system would cool. Similarly if $\Delta H_r < 0$, you have to cool the system to keep the temperature unchanged. So,

- If $\Delta H_r > 0$ we say that the reaction is *endothermic*
- If $\Delta H_r < 0$ we say that the reaction is *exothermic*

Note also that if

for disappearance of 1 mol A and 3 mol B to create 2 mol R

$$A + 3B \rightarrow 2R \ldots \Delta Hr_r = 1000$$

then for

$$2A + 6B \rightarrow 4R \ldots \Delta H_r = 2000$$

and for

$$2R \rightarrow A + 3B \ldots \Delta H_r = -1000$$

this means for 2 mol R disappearing

So ΔH_r depends on how you write the stoichiometry of the reaction.

- For liquids and solids since $\Delta(pV) \cong 0$ we have $\Delta H_r \cong \Delta U_r$
- For ideal gases only, at temperature T and pressure p, by definition we can write

$$\Delta H_r = \Delta U_r + \Delta(pV) \qquad \Delta(nRT)$$

at constant p   *at constant V*

or

$$\Delta H_r = \Delta U_r + (\Delta n)RT \qquad (9\text{–}3)$$

so if there is no change in number of moles during reaction then $\Delta H_r = \Delta U_r$. ΔH_r is usually more useful than ΔU_r, so it is this quantity that is usually tabulated.

C. REACTION NETWORK AT STANDARD CONDITIONS, 25°C AND 1 ATM

In laboratories and factories we usually run reactions at constant pressure, not constant volume. Thus, we should use enthalpies and Eq. 9–2, not internal energies and Eq. 9–1 to represent the energy change involved.

Consider the general energy map for reactant A and product R shown in Figure 9–3, all at 25°C and 1 atm, all written in terms of enthalpy changes. For the same starting and ending materials the enthalpy change is the same, no matter what path you take. Thus,

$$\Delta H_{f1} + \Delta H_{r3} = \Delta H_{f2}$$

gives ΔH_r from enthalpies of formation. Also,

$$\Delta H_{r3} + \Delta H_{c5} = \Delta H_{c4}$$

gives ΔH_r from enthalpies of combustion. In general then,

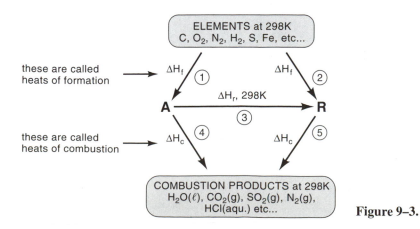

Figure 9–3.

$$\boxed{\Delta H_r = \Sigma \Delta H_f - \Sigma \Delta H_f} \tag{9-4}$$
$$\underset{products}{} \qquad \underset{reactants}{}$$

or

$$\Delta H_r = \Sigma \Delta H_c - \Sigma \Delta H_c \tag{9-5}$$
$$\underset{reactants}{} \qquad \underset{products}{}$$

Thus, heats of formation and combustion can be used to find heats of reaction.

This is most useful because you need not tabulate the ΔH_r for the thousands upon thousands of chemical reactions known or imagined. Just tabulate ΔH_f and/or ΔH_c for each compound of interest. Then you can find ΔH_r for any reaction involving any combination of these compounds.

Why do we want to find ΔH_r for reaction? Because once we know this we can tell whether to heat or to cool the reactor, or whether the reactants have the makings of a good fuel (large negative ΔH_r).

Table 9–1 presents, in shortcut notation, heats of formation for a few selected materials, and Table 9–2 presents heats of combustion, again in shortcut notation, for a few selected materials.

TABLE 9–1. Standard Heats of Formation, $\Delta H_{f,298K}$

The standard heat of formation is the heat of reaction to produce 1 mol of product from its elements, where reactants and product are all at the standard conditions, chosen to be 25°C.

this entry represents the reaction

$$H_2(g) + {}^1\!/_2 O_2(g) \rightarrow H_2O(g) \qquad \Delta H_r = -241.83 \text{ kJ/mol}$$

all at 25°C and 1 atm

$H_2O(g)$	$\Delta H_{f,298} = -241.83$ kJ/mol
$H_2O(\ell)$	-285.84
CO_2	-393.5
CO	-110.5
$C_6H_5CHO(\ell)$, benzaldehyde	-88.83
$Fe_3C(s)$, iron carbide	$+20.92$
HCN(g), hydrogen cyanide	$+130.54$
MgO(s), magnesium oxide	-601.8
$N_2O(g)$, nitrous oxide	$+81.5$
$SO_2(g)$, sulfur dioxide	-296.9

this minus sign means that heat must be removed to produce product at 298 K, hence it represents an exothermic reaction

TABLE 9–2. Standard Heats of Combustion, $\Delta H_{c,298K}$

The standard heat of combustion of a material is the ΔH_r to react (burn or oxidize) one mole of this material to a final state involving $H_2O(\ell)$, CO_2, all materials starting and ending at 25°C.

this entry represents the reaction

$$H_2 + {}^1\!/_2 O_2 \rightarrow H_2O(\ell) \qquad \Delta H_c = \Delta H_r = -285.84 \text{ kJ/mol}$$

$H_2(g)$	$\Delta H_c = $ −285.84 kJ/mol = −141.8MJ/kg	
C(s)	−393.5	−32.8
CO(g)	−283.0	−10.1
$CH_4(g)$, methane	−890.4	−55.5
$C_8H_{18}(\ell)$, octane	−5470.6	−48.0

as another example, this entry represents the reaction

$$C_8H_{18} + 12^1\!/_2 O_2 \rightarrow 8CO_2 + 9H_2O(\ell)$$

$C_{10}H_8$(s), naphthalene	−5157	−40.29
$CH_3 \cdot COOCH_3(\ell)$, methyl acetate	−1595	−21.53
Mg(s), magnesium	−601.8	−24.75
$C_2H_6O_2(\ell)$, ethylene glycol	−1190.3	−19.18
$C_7H_5N_3O_6$(s), trinitrotoluene, TNT	−3435	−15.13

For some larger molecules used as fuels (and also useful for fueling you and me)

Methanol, $CH_3OH(\ell)$ $\Delta H_c = $	−726.55 kJ/mol = −22.7 MJ/kg $\cong$ −5.4 kcal/gm		
Ethanol, $C_2H_5OH(\ell)$	−1366.9	−29.7	−7.1
Glucose, fructose, $C_6H_{12}O_6(\ell)$	−2815.8	−15.6	−3.7
Sucrose, $C_{12}H_{22}O_{11}$(s)	−5643.8	−16.5	−3.9
Starch(s), giant molecules of unknown $\overline{mw}$		−17.5	−4.2
Stearin, animal fat $C_3H_5(C_{17}H_{35}COO)_3$(s)		−39.2	−9.4
Protein(s), a whole family of giant molecules		−21	−5.0

EXAMPLE 9–1.   Reaction in a constant volume bomb

8 gm H_2, 28 gm N_2, and 64 gm O_2 fill a constant volume container at 25°C (Figure 9–4). A spark is passed through the mixture, O_2 and H_2 combine to give H_2O, and the container gets hot. How much heat must be removed to bring the container back to 25°C, at which temperature all the H_2O condenses?

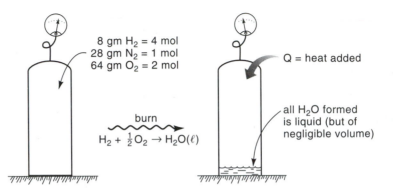

8 gm H_2 = 4 mol
28 gm N_2 = 1 mol
64 gm O_2 = 2 mol

burn
$H_2 + \frac{1}{2}O_2 \rightarrow H_2O(\ell)$

Q = heat added

all H_2O formed is liquid (but of negligible volume)

Figure 9–4.

Solution

First, make a table to keep an accounting of what is disappearing and what is forming.

	Mols before reaction	Mols after reaction
N_2	1	1
H_2	4	0
O_2	2	0
$H_2O(\ell)$	0	4 (liquid)
Total mols gas	7	1

For a constant volume reaction, Eq. 9–1 becomes

$$Q - \cancel{W}_{sh} = \Delta U_r + \cancel{\Delta E_p} + \cancel{\Delta E_k}$$

Let us evaluate ΔU_r. From Table 9–2 we find

$$H_2 + \tfrac{1}{2}O_2 \rightarrow H_2O(\ell) \ldots \Delta H_r = -285\,840 \text{ J}$$

ΔH_r is tabulated; however, ΔU_r is not, but is what is wanted. But by definition, from Eq. 9–2,

$$Q_{added} = \Delta U_r = \Delta H_r - \Delta(p\underline{V})$$

for ideal gases

$$= \Delta H_r - (\Delta n)RT$$

$$= 4(-285\ 840) - (-6)(8.314)(298)$$

$$Q_{added} = \underline{\underline{-1128\ 495\ J}}$$

means that heat is removed

Note: Here 7 moles of gas end up as one mole. In a constant pressure system there would be a big change in volume, work would be done on the system, and Q would differ from what we found above. Thus, in the constant pressure system, whether batch or flow, we have to use ΔH_r to evaluate Q, rather than ΔU_r.

The following examples deal with such systems for which it is proper to use the enthalpy balance.

$$\Delta H + \Delta E_p + \Delta E_k = Q - W_{sh}$$

heat added   *shaft work done by reacting material*

(7–6)

EXAMPLE 9–2.   Humane warfare

Fabulous!! I am told that U.S. Intelligence just developed a most remarkable catalyst guaranteed to make the following reaction go, at room temperature, would you believe!!

$$C_{10}H_8(s) + 6N_2(g) + H_2O(g) \rightarrow 10\ HCN(g) + N_2O(g)$$

naphthalene   *hydrogen cyanide, a most deadly poison*   *nitrous oxide, laughing gas*

Do you know what this means? It represents the ultimate in humane chemical warfare—to have the enemy die while laughing happily.

But before building a chemical reactor to produce this mixture we need to know its heat of reaction. This determines whether we need to cool or heat the reactor. Please calculate this per mole of naphthalene consumed.

Solution

Let us start with a map of the reaction scheme showing the various heats of reaction, from Tables 9–1 and 9–2, and numbering the reactions steps from ①, ②, . . . , to ⑧, as shown in Figure 9–5.

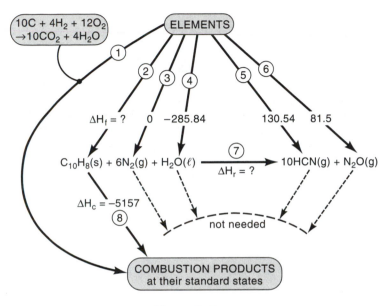

Figure 9–5.

First of all the ΔH_f of naphthalene is not given in Table 9–1, so let us evaluate it from steps ①, ② and ⑧ of the map. Thus,

$$① = ② + ⑧, \text{ or } ② = ① - ⑧$$

Replacing values gives

$$\Delta H_{f,C_{10}H_8} = [10(\Delta H_{f,CO_2}) + 4(\Delta H_{f,H_2O(\ell)})] - \Delta H_{c,C_{10}H_8}$$

and from the values given in Table 9–1,

$$\Delta H_{f,C_{10}H_8} = 10(-393.5) + 4(-285.84) - (-5157)$$

$$= 78.64 \text{ kJ}$$

Now the ΔH_f of all reaction components is known, so make an energy accounting in terms of heats of formation, or

$$② + ③ + ④ + ⑦ = ⑤ + ⑥$$

or

$$⑦ = ⑤ + ⑥ - [② + ③ + ④]$$

This balance is compactly written as Eq. 9–4; replacing values gives

$$\Delta H_r = 10(\Delta H_{f,HCN}) + \Delta H_{f,N_2O} - [\Delta H_{f,C_{10}H_8} + 6\Delta H_{f,N_2} + \Delta H_{f,H_2O(\ell)}]$$

$$= 10(130.54) + 81.5 - [78.64 + 0 + (-285.84)]$$

or

$$\Delta H_r = +1594.10 \text{ kJ/mol } C_{10}H_8 \quad \longleftarrow$$

So this is a highly endothermic reaction, requiring a considerable heat input.

D. REACTIONS AT OTHER THAN STANDARD CONDITIONS

So far we have shown how to find ΔH_r for any reaction at the standard temperature of 298K given ΔH_f or ΔH_c for all the components. Suppose we want to find ΔH_r at other than 298K, say at a temperature T. As a first step, prepare a map as shown in Figure 9–6 and evaluate the energies involved.

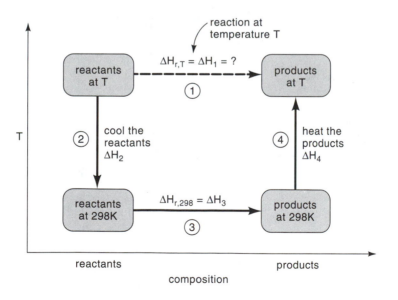

Figure 9–6.

With the same starting and ending materials, the total enthalpy changes will be the same whichever path is taken, so

$$\Delta H_1 = \Delta H_2 + \Delta H_3 + \Delta H_4$$

knowing c_p's and $\Delta H_{r,298}$ will allow you to calculate ΔH_1.

EXAMPLE 9–3.   ΔH$_r$ at various temperatures

From the ΔH$_c$ and ΔH$_f$ tables I've calculated that the standard heat of my gas phase reaction is as follows

$$A + B \rightarrow 2R \quad \ldots \quad \Delta H_{r,298K} = -50\ 000 \text{ J}$$

At 25°C the reaction is strongly exothermic. But that doesn't interest me because I plan to run the reaction at 1025°C. What is the ΔH$_r$ at that temperature, and is the reaction still exothermic at that temperature?

Data   Between 25°C and 1025°C the average c$_p$ values for the various reaction components are

$$\overline{c}_{p,A} = 35 \text{ J/mol·K} \quad \overline{c}_{p,B} = 45 \text{ J/mol·K} \quad \overline{c}_{p,R} = 80 \text{ J/mol·K}$$

Solution

First, prepare a reaction map (Figure 9–7).

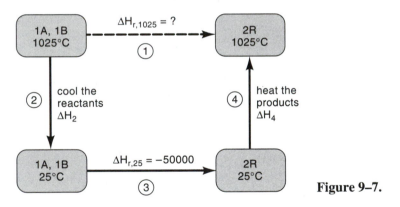

Figure 9–7.

Making an enthalpy balance for 1 mol A, 1 mol B, and 2 mol R gives

$$\Delta H_1 = \Delta H_2 + \Delta H_3 + \Delta H_4$$

$$= (n\overline{c}_p \Delta T)_{\text{reactants}} + \Delta H_{r,25°C} + (n\overline{c}_p \Delta T)_{\text{products}}$$
$$\quad\quad 1A + 1B \quad\quad\quad\quad\quad\quad\quad\quad 2R$$

$$= 1(35)(25 - 1025) + 1(45)(25 - 1025) + (-50\ 000) + 2(80)(1025 - 25)$$

or

$$\Delta H_{r,1025°C} = 30\ 000 \text{ J} \quad \longleftarrow$$

The reaction is $\begin{cases} \text{exothermic at 25°C} \\ \text{endothermic at 1025°C} \end{cases}$

E. EXTENSIONS

There are numerous variations to this simple situation. Let us illustrate the energy map for these.

1. Find ΔH_r when reactants are at T_1, products at T_2, given c_p values and $\Delta H_{r,298K}$. From the map (Figure 9–8) we see that

$$\Delta H_1 = \Delta H_2 + \Delta H_3 + \Delta H_4$$

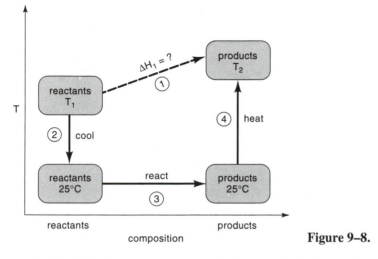

Figure 9–8.

2. Find the flame temperature T_2 for an adiabatic reaction with reactants at T_1, given c_p values and $\Delta H_{r,298K}$. From the map in Figure 9–9 we see that

$$\Delta H_1 = 0 = \Delta H_2 + \Delta H_3 + \Delta H_4$$

from which T_2

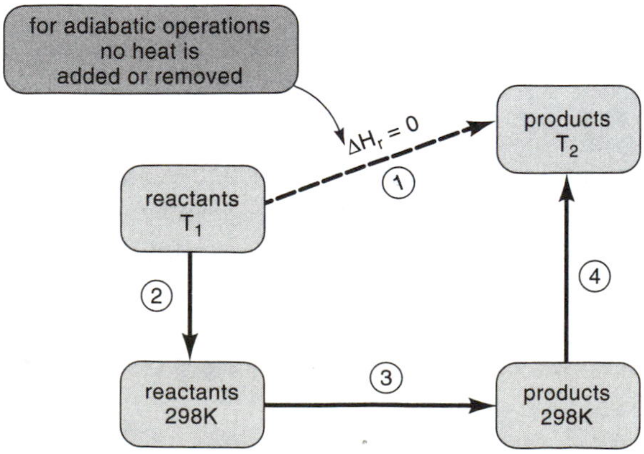

Figure 9–9.

3. Find ΔH_{r1} at T_1 when one of the components, say R, is a gas at T_1, but is a liquid at 298K, given the stoichiometry

$$A(g) \rightarrow R(g)$$

and given c_p values, $\Delta H_{\ell g}$, and $\Delta H_{r,298K}$. From the map in Figure 9–10 we find

$$\Delta H_1 = \Delta H_2 + \Delta H_3 + \Delta H_4 + \Delta H_5 + \Delta H_6$$

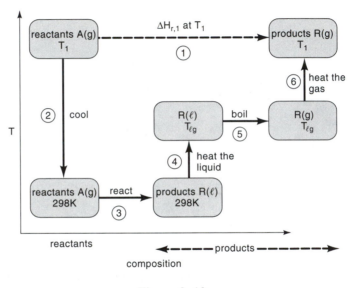

Figure 9–10.

Let us now take a numerical example. As you follow it note that you have to account for the inert material accompanying the reaction components.

EXAMPLE 9–4.   An ugly reaction

One mole of gaseous A, two moles of gaseous B, and one mole of gaseous inert I are introduced into an adiabatic constant pressure variable volume reactor at 25°C. At this temperature reaction would proceed as follows

$$A(g) + B(g) \rightarrow R(\ell) \ldots \text{at } 25°C \tag{9–6}$$

However, the reaction is exothermic and the product heats to 325°C where R is a gas. Find $\Delta H_{r,298K}$ for the reaction of Eq. 9–6.

Data   For A: $c_p(g)$ = 30 J/mol·K

               For B: $c_p(g)$ = 40 J/mol·K

               For R: $c_p(\ell)$ = 60 J/mol·K

                        $c_p(g)$ = 80 J/mol·K

               For I: $c_p(g)$ = 30 J/mol·K

R melts at $T_{s\ell}$ = −25°C

R boils at $T_{\ell g}$ = 125°C

For R: $\Delta H_{\ell g}$ = 10 000 J/mol

Solution

Basis for the calculation: 1 mol A

Let us sketch the reactor (Figure 9–11).

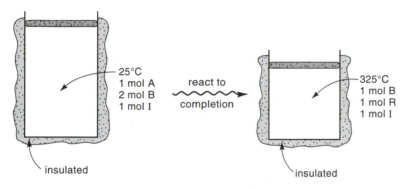

Figure 9–11.

Next, prepare a reaction map (Figure 9–12).

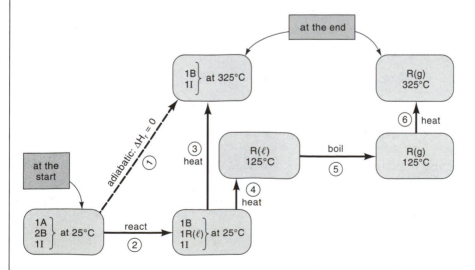

Figure 9–12.

Equating the enthalpy changes for the two paths gives

$$\Delta H_1 = \Delta H_2 + \Delta H_3 + \Delta H_4 + \Delta H_5 + \Delta H_6$$

or

$$0 = \Delta H_{r,298K} + [n_B \Delta H_B (25 \rightarrow 325) + n_I \Delta H_I (25 \rightarrow 325)]$$

$$+ \; n_R \Delta H_{R,\ell} (25 \rightarrow 125) + n_R \Delta H_{\ell g} \text{ (at 125°C)} + n_R \Delta H_{R,g} (125 \rightarrow 325)$$

Replacing values

$$0 = \Delta H_{r,298K} + [1(40)(300) + 1(30)(300)]$$

$$+ \; (1)(60)(100) + (1)(10\,000) + (1)(80)(200)$$

$$= \Delta H_{r,298K} + 12\,000 + 9000 + 6000 + 10\,000 + 16\,000$$

$$\Delta H_{r,298K} = \underline{-53\,000\text{ J}} \text{ for reaction of Eq. 9–6} \quad \longleftarrow$$

Note: This chapter deals with batch reactors. However, Chapter 13 will show that the values obtained from the constant pressure batch systems are the same as for the constant pressure flow reactor. Thus, if the reaction of this example were run in a flow reactor we would get the same numbers as found here (Figure 9–13).

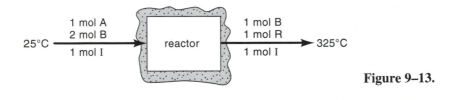

Figure 9–13.

We deal with flow reactor systems in Chapter 13.

F.   ΔU DUE TO MASS CHANGE

Until Einstein performed his magic we had two separate laws, the conservation of mass and the conservation of energy. Einstein's special theory of relativity combined these into a more general law, useful for very high velocity and very large energy changes. Just as originally

- E_k and E_p were related by Galileo
- then heat by Joule and Mayer

- then electricity by Faraday
- then chemical reaction, surface tension, stress-strain, and so on by Van't Hoff, Weber, and the other heroes of thermo

Einstein took the final and ultimate step. He said that every kilogram of mass, no matter what it was, whether bird feathers, bird brains, or iron bars had the same amount of energy. His famous equation $E = mC^2$ in our symbols becomes

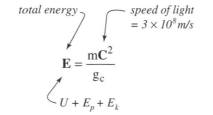

$$E = \frac{mC^2}{g_c}$$

(9–7)

or

$$\Delta E = \frac{(\Delta m)C^2}{g_c}$$

Here we are considering such enormous amounts of energy that we can ignore the difference between E, U, and H. It's like ignoring pennies when dealing with trillions of dollars.

For a delightfully simple presentation of the special theory of relativity, wander through the first 88 pages of L. R. Lieber's *The Einstein Theory of Relativity* (London: Denis Dobson Ltd., 1949).

EXAMPLE 9–5. National energy use

It is estimated that the United States consumes about 2×10^{15} W·hr of electrical energy annually. How many kilograms of matter would have to be destroyed yearly to produce this much energy?

Solution

We are told that $E = 2 \times 10^{15}$ W·hr/year. The mass equivalence, from Eq. 9–7, is given by

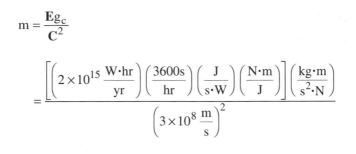

$$m = \frac{Eg_c}{c^2}$$

$$= \frac{\left[\left(2 \times 10^{15} \, \frac{W \cdot hr}{yr}\right)\left(\frac{3600s}{hr}\right)\left(\frac{J}{s \cdot W}\right)\left(\frac{N \cdot m}{J}\right)\right]\left(\frac{kg \cdot m}{s^2 \cdot N}\right)}{\left(3 \times 10^8 \, \frac{m}{s}\right)^2}$$

or

$$m = 80 \, \frac{kg}{yr} \quad \longleftarrow$$

PROBLEMS

1. Find the difference in heat evolved when 1 mole of CH_4 is burned at constant pressure and when it is burned at constant volume.

   Data   Initial pressure = 0.9 bar
   Temperature before and after = 146°C

2. Find the difference in heat evolved when 1 mole of glucose ($C_6H_{12}O_6$) is burned at constant pressure and when it is burned at constant volume.

   Data   Initial pressure = 1.274 bar
   Temperature before and after = 208°C

3. My special home-warming stove can burn either methane or gasoline (assume it to be octane). Now the local energy supermarket sells both of these fuels and they cost the same per kilogram. Which should I buy if I want the most heating per dollar spent?

4. Find the standard heat of combustion at 25°C of 1 kg of liquid benzaldehyde.

5. Find $ΔH_f$ of methyl acetate from its heat of combustion.

6. Ho! Ho! Ho! Last week at Murphy's Tavern I really pulled a fast one. I bought the patent rights to a fantastic process for producing pure shiny magnesium metal and pure methyl acetate from cheap starting materials. The stoichiometry is as follows:

$$2\,MgO + 3C + 3H_2 \xrightarrow[\text{room temperature}]{\text{special secret catalyst}} 2Mg + CH_3{\cdot}COOCH_3$$

↙ *very cheap*   ↙ *expensive*

Unfortunately the fellow who sold me the process and its catalyst didn't tell me whether I would have to cool or heat the reactor . . . and he doesn't seem to be at the address he gave me. So please tell me for each ton of magnesium produced how much heat must be added or removed (in kJ) from the reactor I plan to build. Hurry! I can't wait.

7. That crook at Murphy's really took me to the cleaners. But that's all past. My new deal is completely different. Why, my partner is a real professor and he has two Ph.D.'s (he showed me his diplomas). We have a partnership, 50-50. I supply the money for the factory. He'll supply the knowhow and will build it. The plan is ingenious—to transform scrap iron into useful products. The reaction is as follows:

$$3\,Fe + CH_4 \rightarrow Fe_3C + 2H_2$$

     ↑   ↑   ↑   ↑

   scrap   *waste*   *expensive*   *useful*

   iron   *gas*   *solid*   *fuel*

We need to know how much heat is involved in the reaction. He, the professor, says that it is beneath his dignity to solve so simple a problem, so he asked me. I'm embarrassed to admit that I can't do it, so I turn to you. Please, please could you calculate $\Delta H_{r,298K}$ for me?

8. **(a)** What is the internal energy change ΔU_r at 298K for the reaction of Example 9–4?

    (b) Suppose the reaction of Example 4 was run in an adiabatic constant volume batch reactor (not constant pressure), would the final temperature be higher, lower, or the same as 325°C?

    Note: No need to evaluate the temperature.

9. *Science News*, *143*, 410 (June 26, 1993) reports that a newcomer in our solar system, the newly discovered comet Shoemaker-Levy, passed too close to Jupiter, was trapped by Jupiter's gravitational field, and now cannot escape the planet's tug. In fact, astronomers calculate that on about July 22, 1994, it will return and crash into Jupiter in a spectacular and fiery death, which could release the energy equivalent to the combustion of 1 billion megatons of TNT. If this would have happened to the Earth, not

Jupiter, you'd be able to forget about thermo. It is estimated that the comet has a diameter of 10 km, a density of 5400 kg/m^3, and should crash into Jupiter at 60 km/s.

Question: Is *Science News* a British or American magazine?

10. Modern oil tankers displace 300 000 tons and travel at 45 km/hr. That is quite a lot of kinetic energy. How much mass is this equivalent to?

11. Find the difference in mass when hydrogen and oxygen combine at 25°C to produce 1 kg of liquid water.

12. The Corvallis *Gazette-Times* (May 31, 1980) had an article on the Mt. St. Helens volcanic eruption that said in part:

> . . . the mixture burst like an immense nuclear land mine, catapulting at least a cubic mile of rock at least 70,000 feet into the skies . . .

Time (June 2, 1980) stated that the energy release was equivalent to 500 Hiroshima-type atom bombs, each rated at about 25 000 tons of TNT. Let the density of ejected mountain rock be $\rho_s = 2500$ kg/m^3. This value of 500 atom bombs seems to be awfully large—please check *Time*'s estimate.

13. V. L. Sharpton and colleagues (*Science, 261*, 1564 (September 17, 1993)) forwarded the seventh 1993 version of how the dinosaurs were wiped out about 65 million years ago. They speculated that at that time an object about 12 km in diameter came crashing to Earth just off the tip of the Yucatan peninsula. Its impact supposedly produced the largest explosion to rock our Earth since life began, leaving a crater 300 km in diameter. They calculated the colossal energy release to equal the detonation of 300 million hydrogen bombs, each some 70 times more powerful than the Hiroshima atomic bomb, itself equivalent to the combustion of about 25 000 tons of TNT.

(a) At what velocity did it crash into the earth?

(b) Do you agree with the reported energy release (300 million H bombs, etc.)?

Estimate the density of this cosmic visitor to be $\rho_s = 5400$ kg/m^3.

14. How long can we operate on 1 gigawatt fusion power plant per ton of hydrogen transformed into helium, according to the reaction

$$2H_2 \rightarrow He$$

Atomic masses: H: 1.008 gm,   He: 4.003 gm

15. Nuclear fusion is one of man's most exciting hopes for cheap inexhaustible energy, and the reaction of deuterium (D) and tritium (T), both forms of heavy hydrogen, is probably the easiest reaction to make go. It also has the highest energy yield of the many reactions being considered today. The starting materials are also not that difficult to obtain. The reaction is

$$D + T \rightarrow {}_2He^4 \rightarrow {}_0n^1$$

How many kg of D and how many kg of T must fuse/day for a production rate of 1GW of energy output (roughly equivalent to a good-sized electrical power plant)?

Data   The masses of the various components are

For neutrons: ${}_0n^1$ = 1.008 665 gm/gm atom
For ordinary helium: ${}_2He^4$ = 4.002 60 gm/gm atom
For deuterium: $D = {}_1H^2$ = 2.0140 gm/gm atom
For tritium: $T = {}_1H^3$ = 3.016 05 gm/gm atom

16. How long can we run a gigawatt power plant by fission of 1 ton (metric) of uranium fuel? Note that each gram of uranium gives 0.999 085 gm of fission products.

17. When uranium-235 is bombarded by neutrons going at the right speed, it breaks down into smaller particles with the release of much energy. This process is used for electric power generation today. Somewhat simplified, the fission reaction can be represented as follows:

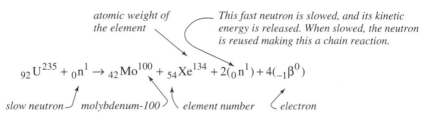

$$\underset{\text{slow neutron}}{{}_{92}U^{235} + {}_0n^1} \rightarrow \underset{\text{molybdenum-100}}{{}_{42}Mo^{100}} + {}_{54}Xe^{134} + 2({}_0n^1) + 4({}_{-1}\beta^0)$$

atomic weight of the element — *This fast neutron is slowed, and its kinetic energy is released. When slowed, the neutron is reused making this a chain reaction.*

slow neutron   *molybdenum-100*   *element number*   *electron*

Note that this is a chain reaction. In terms of energy output how many metric tons of coal (assumed to be pure carbon) are equivalent to burning 1 kg of uranium-235?

Data

Atomic mass of $_{92}U^{235} = 235.112\ 40$ gm, $_{54}Xe^{134} = 133.965\ 17$ gm,

$$_0n^1 = 1.008\ 98 \text{ gm,}   _{-1}\beta^0 = 0.000\ 55 \text{ gm}$$

$$_{42}Mo^{100} = 99.946\ 86 \text{ gm}$$

these are also called atomic weights or atomic mass units, amu

18. We plan to send a space probe to the star α-Centauri, which is about 4 light years away. We will use an advanced "photon" engine to accelerate our craft to three-fifths the speed of light, the speed for the journey.

   (a) How much energy must be expended to get the spacecraft to this speed? The mass of spacecraft will be 100 kg.

   (b) What conversion of mass to energy is this equivalent to?

CHAPTER 10

ENERGY RESERVES AND USE

This chapter presents figures and tables aimed at giving the reader an appreciation of the various aspects of energy in our lives.

Power Producers and Users	Power Involved
Lifting a mosquito at 1 cm/s	$1 \text{ erg} = 10^{-7} \text{ W} = 10^{-10} \text{ kW}$
A fly doing one pushup/s	
Cricket's chirps	$10^{-3} \text{ W} = 10^{-6} \text{ kW}$
Pumping human heart	1.5 W
Burning match	$10 \text{ W} = 10^{-2} \text{ kW}$
Electrical output of a 1 m² solar cell at 10% efficiency	$100 \text{ W} = 0.1 \text{ kW}$
Bright light bulb	$100 \text{ W} = 0.1 \text{ kW}$
Man hard at work (for 1 hr)	0.1 kW
(long term, a slave)	0.02 kW
Draft horse	1 kW
Portable floor heater	1.5 kW
Compact automobile	100 kW
Queen Elizabeth, giant ocean liner	200 000 kW
Boeing 747, passenger jet, cruising	250 000 kW
One large coal fired power plant	$1 \times 10^6 \text{ kW} = 1 \text{ GW}$
Niagara Falls, hydroelectric plant	$2 \times 10^6 \text{ kW} = 2 \text{ GW}$
Space shuttle orbiter (3 engines) plus its 2 solid boosters, at take off	14 GW
All electric power plants worldwide	$10^9 \text{ kW} = 1000 \text{ GW}$
U.S. automobiles, if all used at the same time (100 million)	$10 \times 10^9 \text{ kW} = 10\,000 \text{ GW}$
Man's total use today	$4 \times 10^9 \text{ kW} = 4000 \text{ GW}$
$1\mathbf{Q}/\text{s} = 10^{18} \text{ Btu/s}$	$30 \times 10^9 \text{ kW} = 30\,000 \text{ GW}$

A.   CUMULATIVE ENERGY USE

We need a useful measure for a large amount of energy. For this purpose the unit **Q** was defined

$$\mathbf{Q} = 10^{18} \text{ Btu/s} = 3.35 \times 10^{10} \text{ kW}$$

$$= 33 \text{ million MW}$$

$$= 33\,000 \text{ GW}$$

then the cumulative use of energy by man is shown in Figure 10–1.

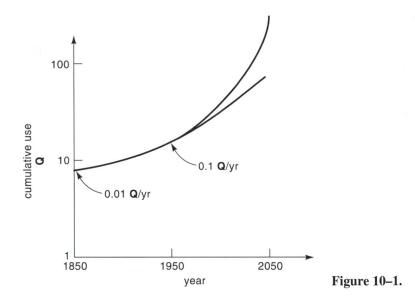

Figure 10–1.

B.   WORLD ENERGY USE TODAY

Source	Percentage	
Coal	46%	84% fossil . . . created and
Oil and natural gas	38%	stored in the past
Farm wastes	10%	
Wood	4%	
Other (hydro, wind, nuclear)	2%	

C. WORLD RESERVES OF ENERGY

	Known Reserves	Potential Reserves	Total Reserves
Coal and peat	25 Q	53 Q	78 Q
Liquid hydrocarbons	2 Q	28 Q	30 Q
Natural gas	0.7 Q	7.6 Q	8.3 Q
Oil shale	—	13 Q	13 Q
	27 Q	102 Q	129 Q

this will last about 80 years

Cost of Energy	¢/kW·hr
Fuel oil	0.6
Natural gas, wood, coal	0.8
Gasoline	1.2
Electricity	3
Sugar, bread, butter	15
Martini, whiskey	750
Flashlight	1500
Caviar	1850

The above tables show that one day's hard work by man (10 hours) produces about $1/2$ kW·hr of useful work. This is worth just a few cents. This is why slavery went out of business.

D. ENERGY OF FOOD (HEAT OF COMBUSTION)

	Δh_r, [kJ/gm]	Food calories/gm (or 1000 cal/gm)
Hydrogen	142	34
Methane	55	13
Gasoline, kerosine, fats	40, or 38–46	9–11
Carbon, alcohols	30	7
Proteins	20, or 19–23	4.5–5.5
Carbohydrates (sugar, starch, etc.)	16, or 15–17	3.6–4.2

A normal person gets by on about 8000 kJ/day. This amounts to about 200 gm of fat/day.

E. ENERGY NEEDS OF HUMANS

	ΔH_r = energy needed
Basal metabolism, just enough to survive	4000 kJ/day
Teacher, student, and office worker	8000 kJ/day
Bike racer, logger, iron man	20 000 kJ/day

Noting that some foods contain lots of water, others don't, we find that a normal person can get by on 200 gm chocolate/day, or 400 gm of steak/day, or 3.5 kg of broccoli/day.

F. ENERGY STORAGE IN HUMANS

A normal individual has about 300 000 kJ of energy reserves on his or her body, to use in case of emergency.

Conversion of food to fat and muscle is about 10% efficient. The other 90% is used to keep the individual warm. Thus,

$$100 \text{ kg of corn or grass} \rightarrow 10 \text{ kg of cow} \rightarrow 1 \text{ kg of man}$$

If you cut out one step in the conversion,

$$100 \text{ kg of corn} \rightarrow 10 \text{ kg of man}$$

then your species can multiply. That's why there are many gazelles and few tigers. And when food becomes scarce we all will have to become vegetarians.

Note: The values presented in this chapter were collected from various sources. However, most came from that delightful popularly written thermo book: *Order and Chaos* by S. W. Angrist and L. G. Hepler (New York: Basic Books, 1967). Try to borrow or steal a copy. It's a treasure.

EXAMPLE 10–1.   Mercy, Russian style

Catherine the Great ruled Russia with an iron hand at the time of the American Revolution. She was a no-nonsense ruler who tired of her husband, Tsar Paul, and had him killed. She also ordered neck stretching jobs on many of those courtiers who displeased her or could not satisfy her desires.

In her whole reign, only once did she show mercy and compassion. Curiously, it was for an uncouth, foul-mouthed, unrepentant revolutionary named Igor Dimitrievich Grebenschikoff. To the amazement of all, she commuted his death sentence to imprisonment for as long as she reigned, but on a diet of water and a 1 kg loaf of standard prison-baked black bread per week, plus one cube of sugar (30 gm) each Christmas. After her death he was to be set free to harass the next ruler of Russia. Do you think that Igor returned to his revolutionary activities twelve years later, at the end of Catherine's rule, or do you think that he was reformed?

Data Russian prison-grade black bread consisted of 10% protein, 80% carbohydrate, and 10% straw and sawdust.

Solution

Let's see how he would fare on that diet, because a loaf a week seems pretty skimpy. The food energy per kilogram loaf of bread is then

(20 kJ/gm) (100 gm protein) + (16 kJ/gm) (800 gm carbohydrates) = 14 800 kJ/loaf

So Igor's energy intake is

(14 800 kJ/loaf) (1 loaf/7 days) = 2114 kJ/day

But the fuel his body needs to just stay alive = 4000 kJ/day
So he must draw on his body's reserves by 4000 − 2114 = 1886 kJ/day
But the average human reserve of energy, in fat, muscles, etc. = 300 000 kJ.
Therefore, Igor runs out of reserves after

$$\frac{300\ 000\ \text{kJ}}{1886\ \text{kJ/day}} = 159 \text{ days, or } \underline{\text{less than 6 months}}$$

These are rough figures, because we don't know whether Igor was a thin, wiry chap, or a roly-poly. But these calculations show that no way can he outlive the Empress. It will be a slow agonizing death for him, for sure.

PROBLEMS

1. On the average I consume 8000 kJ/day of food. But I am a bit flabby, so as soon as the term is over I will diet on 4000 kJ/day. Boy oh boy, the fat will just melt away and in no time I'll lose 5 kg. I wonder how long "no time" will be? In other words, estimate how long I must diet to lose those 5 kg of fat.

2. *Chemical and Engineering News* (September 23, 1985), reports in its "Department of Obscure Information" that "the energy content of 240 bottles of table wine is about 1 million BTU." I think that this is rubbish, nonsense. Would you please check this for me?

   Data   A normal sized bottle of wine contains 0.75 lit of fluid.

3. How many liters of pure oxygen at 25°C and 1 atm do I use up each day in breathing if my food intake is about 8000 kJ/day? As a personal note, I love candy bars and sweets and practically live on that stuff.

4. On a can of diet cola, costing 60¢, I read that it contains one calorie of food energy. How much does a kW·hr of cola energy cost, and how does this compare with the cost of caviar?

5. If we could harness the release of body heat of the population of New York City (say 10 M people) roughly what rate of energy dissipation would this represent, measured in terms of giant 1 GW nuclear power plants?

6. We must plan for the future. Time will soon be upon us when gasoline will cost $3/gal. We must find an alternative liquid fuel. Alcohol from the fermentation of vegetable matter is one strong possibility.

   (a) How many kJ of energy can we get per dollar's worth of gasoline?

   (b) How much should ethanol cost/gallon if it is to compete with gasoline as a fuel?

   Data   Assume the same efficiency of utilization of energy.
   Assume gasoline to be pure octane: $\rho = 700 \text{ kg/m}^3$
   Assume alcohol to be pure ethanol: $\rho = 790 \text{ kg/m}^3$

7. The candy salesman told me that Hershey milk chocolates (plus one vitamin pill per day) are a complete food. That was good news to me, so I rushed out and bought a truckload of these candy bars, all at a fantastic discount. They should last me quite a while, maybe even into the next century. But to determine this I must know how many to eat each day. You see, I am an 8 million joule per day man. So please calculate this for me.

   Data   The labels on the Hershey bars show that they are 1.4 oz candy bars that consist of 3 gm protein, 23 gm carbohydrate, and 13 gm fat.

8. The California grey whale, a giant mammal weighing about 40 tons fully grown, spends its summers in the Bering Sea off Alaska where for six months it eats, eats, eats, and fattens up. Then it heads south to the waters of Mexico to breed. It mates one winter, and gives birth the next winter. This

represents the longest annual migration of any mammal on earth. In any case, during this 6-month period the whale does not eat but lives off its stored fat, and loses about 5 tons of mass. What is its activity rate compared to man?

9. This has been a tough term and what I'd really like to do is just go to my study, take it easy and snack on sweets and candy bars. Of course I won't overdo it . . . just 8000 kJ/day of that yummy stuff. My study is a warm (25°C), comfortable, soundproof, airtight room about 3 m × 4 m × 2.5 m high. I'll relax and snack and relax some more until, until . . . well until I've used up one-fourth of the oxygen in the room. I wonder how long that will take. Would you please figure it out for me? Maybe after a good rest I'll think of dieting.

10. If my car breaks down and leaves me stranded and starving I wonder whether, as a last resort, I can drink the antifreeze fluid (practically pure ethylene glycol). Kilogram for kilogram how much food energy would this fluid give compared to sugar? However, before I leave on my travels I'd better check whether ethylene glycol is toxic.

11. Payless Drug Stores' sports department has a brand new item, a really versatile camp stove that burns either gasoline or sugar. I want to know which fuel I should burn, so would you let me know? More precisely:

    (a) Determine the heat release in MJ per dollar of gasoline burned.

    (b) Determine the heat release in MJ per dollar of sugar burned.

    > **Data** Sugar is pure sucrose, $C_{12}H_{22}O_{11}$.
    > Today sugar costs $1.99/5-pound bag at the grocery store.
    > Assume gasoline is pure octane.
    > Density = 703 kg/m^3 for gasoline.
    > Today 100 octane gasoline costs $1.399/gallon.

12. *Chemical and Engineering News*, pg. 90 (October 18, 1993) reports that the average adult inhales 440 ft^3 of air per day. Check to see whether this figure is reasonable.

    > **Data** Schmidt-Nielsen, *Scaling: Why Is Animal Size So Important?* (Cambridge University Press, 1984), pg. 104, tells that the oxygen absorbed in the lung is 3.1% of the air intake.

13. Some scientists propose that Earth's early atmosphere consisted primarily of CO_2, N_2, and water vapor, with negligible oxygen. They also assume that

oxygen was only formed when plant life invaded Earth's surface, flourished, died, decayed, and was buried underground. Crudely, we may represent this as follows:

$$CO_2 \xrightarrow{\text{sunlight}} \text{plant life} \xrightarrow[\text{burial}]{\text{decay}} C + O_2 \uparrow$$

coal, oil, natural gas

A negligible fraction of the fixed carbon is presently stored in vegetation today; most is locked away and buried as coal, shale, and hydrocarbon deposits. Based on this picture of our past, and assuming that all buried organic material is carbon alone, can you estimate what fraction of the underground stored coal and hydrocarbon deposits have been found today?

> **Data**   Earth circumference = 40 000 km
>
> Mole fraction of oxygen in the atmosphere = 0.21.

14. As an old-time mountaineer, on Friday, February 2nd, I will join the OSU Mountain Club to climb Mount Hood from Timberline Lodge (1829 m) to very top (3427 m above sea level). Alcohol with a bit of melted snow will fuel my trip. If the transformation of fuel to climbing efficiency is 20% (crude estimate) how many $1/2$ liter bottles of pure alcohol would I need to take along?

> **Data**   With pack, camera, and fuel I weigh 90 kg.

15. For our *Reduce and Be Crabby* cookbook we want to report the energy value of sugar, which is practically pure sucrose, $C_{12}H_{22}O_{11}$. Thermodynamic tables give the following heats of formation:

> $\Delta H_f = -2218$ kJ/mol sucrose
> $\Delta H_f = -393.5$ kJ/mol CO_2
> $\Delta H_f = -285.4$ kJ/mol H_2O (ℓ)

(a) Only use the above heats of formation data to calculate the energy value of a teaspoon of sugar.

(b) Check your answer with the value given in cookbooks.

> **Note:**   In American cookbooks:
>
>   1 tablespoon = 1/16 cup = 1/32 pint = 3 teaspoons = 12 gm of white sugar

In other countries different measures are used. For example, in English cookbooks:

   1 tablespoon = 4 teaspoons.

Aside: You may be interested to know that the Sugar Association and all the *Eat and Be Happily Fat* cookbooks report that 1 teaspoon of sugar supplies 15 Cal of food energy.

16. *Bird Flight Performance* by Pennycuick (Oxford: Oxford University Press, 1989) tells that a 0.31 kg bird with a 0.6 m wingspan consumes 34 watts when it flies at its optimum speed of 15.1 m/s. It also tells that this is an energy equivalent of 0.058 gm of fat/km, or about 34 000 miles/gallon of gasoline. Please check these last two figures.

Data Density of gasoline: $\rho = 700$ kg/m^3

17. In Chapter 6 we read that the kinetic energy of 1 gm of material traveling at $v = 0.99C$ is equivalent to the chemical energy in 17 000 m^3 of gasoline. Please verify this. Assume that gasoline consists of pure octane, with density $\rho = 703$ kg/m^3.

CHAPTER 11

THE IDEAL GAS
AND THE FIRST LAW

Let us see what interesting and useful treasures appear when we apply the first law to ideal gases. To remind you, the behavior of all gases at not too high a pressure can reasonably be predicted and accounted for by the ideal gas law relationship

$$\underset{\underset{m^3}{\nearrow}}{\overset{\overset{Pa}{\downarrow}}{p}}\, V = \underset{\underset{8.314\ J/mol\cdot K}{\nwarrow}}{\overset{\overset{mol}{\downarrow}\ \overset{K}{\swarrow}}{n\,R\,T}} \quad \text{or} \quad p\,\overset{\overset{per\ mole}{\downarrow}}{v} = R\,T \tag{11-1}$$

Let us describe a very simple and famous experiment that links the ideal gas to the first law and tells us the energy properties of these gases.

A. THE JOULE EXPERIMENT

Two interconnected flasks are immersed in water as shown in Figure 11–1. Let the flasks and their contents be the system and let the water be the surroundings. One flask contains a gas at high pressure, the other is evacuated. All is isothermal at T_1. The stopcock is then opened, and the pressure in the two flasks equalizes. What Joule found is that for an ideal gas the temperature of the system and surroundings did not change, thus $T_1 = T_2$.

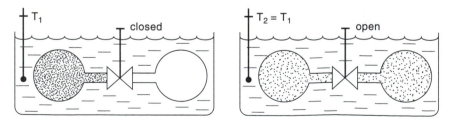

Figure 11–1. The Joule experiment and its important finding that $T_2 = T_1$.

That's all there is to this experiment. I wonder whether you could pull a significant finding from this experiment. Try. Give up? Let us continue. Consider the two-flask system with its gas and let us see what the first law tells us. From the closed system expression, Eq. 3–2 or 7–5 we write

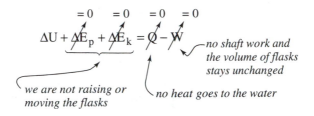

Thus, the first law tells that the internal energy of the gas remains unchanged, $\Delta U = 0$.

But Joule also found that $\Delta T = 0$, even though p and V do change. This means that $U \neq f(p,V)$, but that $U = f(T)$ alone. In symbols then

$$\boxed{\text{For a mole of ideal gas} \quad \Delta u = c_v \Delta T \quad \text{... independent of p, v}}$$

$$\text{more generally} \quad \Delta u = \int_{T_1}^{T_2} c_v \, dT$$

(11–2)

Next consider the enthalpy change of an ideal gas. By definition for one mole of gas

$$h = u + pv \qquad (7–1)$$

and for a change from state 1 to state 2

$$\Delta h = \Delta u + \Delta(pv)$$

$$= \Delta u + \Delta(RT)$$

with constant c_v

$$\Delta h = c_v \Delta T + R \, \Delta T$$

function of T alone, independent of p or V, from Eq. 11–2

(11–3)

Also by definition

$$\Delta h = c_p \Delta T \qquad (11–4)$$

Summarizing the above findings we have

For one mole of ideal gas

$$\left.\begin{array}{l} \Delta u = c_v \Delta T \\[4pt] \Delta h = c_p \Delta T \\[4pt] c_p = c_v + R \end{array}\right\} \text{ independent of p, v} \tag{11-5}$$

from Eq. 11–3 and Eq. 11–4

If c_v and c_p vary with temperature we write more generally

$$\Delta u = \int_{T_1}^{T_2} c_v\, dT \quad \text{and} \quad \Delta h = \int_{T_1}^{T_2} c_p\, dT \tag{11-6}$$

We will see that these are very useful relationships.

A final note on ideal gases: Using the kelvin temperature scale, as in Eq. 11–1, we see that when $T \to 0$, so does pV. Then, from Eq. 7–1 we see that $h = u$ at zero kelvin. Thus, at any temperature T

$$u = \int_{0}^{T} c_v\, dT \quad \text{and} \quad h = \int_{0}^{T} c_p\, dT$$

and for constant c_p or c_v

$$\left.\begin{array}{l} u = c_v T \\[4pt] h = c_p T \end{array}\right\} \quad \text{T in kelvin,} \quad [\text{J/mol·K}] \tag{11-7}$$

B. EQUATIONS REPRESENTING CHANGES IN A BATCH OF IDEAL GAS

This section develops expressions for the work and heat involved in various changes of an ideal gas. But first—a few words about the concept of work.

Imagine a high pressure gas pushing back a piston and expanding in a cylinder. It does work, both mechanical and pV work (see Chapter 7). If this occurs without friction or heat loss, in effect, adiabatically and reversibly, then

$$W_{\substack{\text{total done by} \\ \text{expanding gas}}} = W_{\substack{\text{total received by} \\ \text{surroundings}}} \tag{11-8}$$

But suppose the piston is rusty and not well lubricated so that it grinds on the walls of the cylinder as it moves outward, generating friction and heat. Here the surroundings will receive less total work than that done by gas, so

$$W_{\substack{\text{done by gas}}} > W_{\substack{\text{received by} \\ \text{surroundings}}} \tag{11-9}$$

In this section all the work and heat expression are calculated by assuming frictionless, or reversible, operations. We indicate this by attaching the subscript "rev" to the work and heat terms, or w_{rev} and q_{rev} in the expressions that follow. Also note that we are considering the total work produced, not just shaft or pV work.

To derive these expressions we apply the first law and the energy relations developed earlier in this chapter. Per mole of a batch of ideal gas the equations we use are

$$
\begin{array}{c}
\Delta_u = c_v \Delta T \\
pv = RT \quad \Delta h = c_p \Delta T \\
c_p = c_v + R
\end{array}
\quad
\begin{array}{l}
\text{and with } \Delta e_p = 0 \text{ and } \Delta e_k = 0 \text{ we get} \\
\Delta u = q_{rev} - w_{rev} \dots \text{for a closed system}
\end{array}
\qquad (11\text{--}10)
$$

Now on to the various changes, remembering that w_{rev} represents both the shaft and pV work.

1. Constant Volume Process

$$
v_1 = v_2 \quad \text{and} \quad \frac{p_1}{T_1} = \frac{p_2}{T_2}
$$

$$
w_{rev} = \int p\, dv = 0
$$

$$
q_{rev} = \Delta u + w_{rev} = c_v \Delta T + 0 = c_v \Delta T \quad \left[\frac{J}{mol} \right]
$$

$$(11\text{--}11)$$

2. Constant Pressure Process

$$
p_1 = p_2 \quad \text{and} \quad \frac{v_1}{T_1} = \frac{v_2}{T_2}
$$

$$
w_{rev} = \int p\, dv = p(v_2 - v_1) = p_1 v_1 \left(\frac{T_2}{T_1} - 1 \right) = \frac{p_1 v_1}{T_1}(T_2 - T_1) = R\Delta T
$$

$$\text{constant} \qquad \frac{v_1 T_2}{T_1}$$

$$
q_{rev} = \Delta u + w_{rev} = c_v \Delta T + R\Delta T = c_p \Delta T \quad \left[\frac{J}{mol} \right]
$$

$$(11\text{--}12)$$

3.   Constant Temperature Process

$$T_1 = T_2 \quad \text{and} \quad p_1 v_1 = p_2 v_2$$

$$\Delta u = 0 \quad \text{because T is constant}$$

$$\Delta h = \overset{= 0}{\cancel{\Delta u}} + \overset{= 0}{\cancel{\Delta (pv)}} = 0 \tag{11-13}$$

$$q_{rev} = w_{rev} = \int p \, dv = \int \frac{RT}{v} \, dv$$

$$= RT \, \ell n \frac{v_2}{v_1} = RT \, \ell n \frac{p_1}{p_2} \qquad \left[\frac{J}{mol} \right]$$

4.   Adiabatic (q = 0) Reversible Process, with Constant c_v

For an adiabatic reversible expansion or compression p, v, and T all change, so let us write the changes in the system in differential form. From the first law

$$du = \overset{= 0}{\cancel{dq}}_{rev} - dw_{rev} = -p \, dv \qquad \left[\frac{J}{mol} \right] \tag{11-14}$$

$$\underset{c_v dT}{\curvearrowleft} \qquad \qquad \underset{\frac{RT}{v} dv}{\curvearrowleft}$$

Separating and integrating gives

$$\int_{T_1}^{T_2} \frac{dT}{T} = -\frac{R}{c_v} \int_{v_1}^{v_2} \frac{dv}{v}$$

Assuming a constant c_v, hence constant c_p, we introduce symbol k as

$$k = \frac{c_p}{c_v} = 1 + \frac{R}{c_v} = \text{constant} \quad \begin{cases} k \cong 1.67 & \text{for monatomic gases} \\ k \cong 1.40 & \text{for diatomic gases} \\ k \cong 1.32 & \text{for larger molecules} \end{cases} \tag{11-15}$$

Then using k, the above equation becomes, on integration,

$$\ell n \frac{T_2}{T_1} = -(k-1) \, \ell n \frac{v_2}{v_1}$$

To summarize

$$\frac{T_2}{T_1} = \left(\frac{v_1}{v_2}\right)^{k-1}$$

$$\frac{T_2}{T_1} = \left(\frac{p_2}{p_1}\right)^{(k-1)/k} \qquad (11\text{--}16)$$

$$\frac{p_2}{p_1} = \left(\frac{v_1}{v_2}\right)^{k}, \quad or \ \ pv^{k} = constant$$

For ideal adiabatic reversible batch processes, with $\Delta e_p = 0$ and $\Delta e_k = 0$, we have

$$q_{rev} = 0$$

$$w_{rev} = \int p \, dv = \int_{v_1}^{v_2} \frac{constant}{v^{k}} \, dv$$

$$= \frac{p_1 v_1}{k-1}\left[1 - \left(\frac{p_2}{p_1}\right)^{(k-1)/k}\right] = \frac{RT_1}{k-1}\left[1 - \left(\frac{p_2}{p_1}\right)^{(k-1)/k}\right] \qquad (11\text{--}17a)$$

and from Eq. 11–14

$$w_{rev} = -\Delta u = -c_v(T_2 - T_1) = -\frac{R}{k-1}(T_2 - T_1) = -\frac{p_2 v_2 - p_1 v_1}{k-1} \ \left[\frac{J}{mol}\right] \qquad (11\text{--}17b)$$

Finally, note that when an ideal gas goes from state 1 to state 2, whether reversibly or not, then all expressions that only involve state properties—T, v, p, c_p, u, and h—always hold. Thus, if you give me T_1, p_1, v_1, T_2, p_2, and c_p, I will give you v_2, Δu, and Δh.

However, the expressions in this chapter that contain q and/or w only hold rigorously for reversible processes. For real processes the calculated work must be modified by some efficiency factor η, where $0 < \eta < 1$.

Thus, when a batch of ideal gas goes from state 1 to state 2

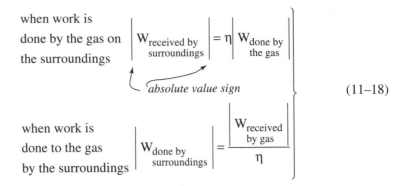

when work is
done by the gas on
the surroundings

$$\left| W_{\substack{received\ by \\ surroundings}} \right| = \eta \left| W_{\substack{done\ by \\ the\ gas}} \right|$$

absolute value sign

$(11\text{--}18)$

when work is
done to the gas
by the surroundings

$$\left| W_{\substack{done\ by \\ surroundings}} \right| = \frac{\left| W_{\substack{received \\ by\ gas}} \right|}{\eta}$$

For nonideal gases the expressions obtained are much more complicated.

C. COMPRESSION AND EXPANSION PROCESSES IN PRACTICE

In general, the equations for such systems can get very complicated, so we would like to use, whenever possible, the simple equation forms derived above. This means adopting what is called a *polytropic process*, represented by

$$p\, v^{\gamma} = constant \tag{11--19}$$

where γ is a constant that takes the place of $k = c_p/c_v$ (Fig. 11–2).

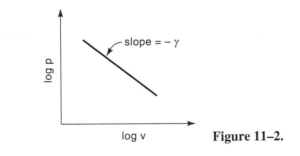

slope $= -\gamma$

log p

log v

Figure 11–2.

For an isobaric (constant pressure) process $\gamma = 0$, and $q \neq 0$
For an isothermal process $\gamma = 1$, and $q \neq 0$
For an adiabatic reversible process $\gamma = k$, and $q = 0$
For a polytropic process usually $1 < \gamma < k$

For polytropic processes integration of Eq. 11–19 gives Eqs. 11–16 and 11–17a, but with k replaced by γ. Thus

$$w_{batch,poly} = \int p \, dv = \int \frac{constant}{v^\gamma} \, dv$$

$$= \frac{p_1 v_1}{\gamma - 1} \left[1 - \left(\frac{p_2}{p_1} \right)^{(\gamma-1)/\gamma} \right] = \frac{RT_1}{\gamma - 1} \left[1 - \left(\frac{p_2}{p_1} \right)^{(\gamma-1)/\gamma} \right] \qquad (11\text{–}20a)$$

and Eq. 11–17b becomes

$$w_{batch,poly} = -\frac{p_2 v_2 - p_1 v_1}{\gamma - 1} = -\frac{R}{\gamma - 1}(T_2 - T_1) = -\frac{k-1}{\gamma - 1}(\Delta u) \qquad \left[\frac{J}{mol} \right] \quad (11\text{–}20b)$$

D.   FLOW PROCESSES FOR IDEAL GASES

The necessary equations for compressors, pumps, turbines, and other machinery using gases at not too high pressure are given in Chapter 13.

E.   EXAMPLES OF VARIOUS CHANGES IN A BATCH OF IDEAL GAS

Consider various possible changes experienced by a tank or cylinder of ideal gas. Let the original temperature be T_1 and the final temperature T_2.

EXAMPLE 11–1.   Slow leak from an insulated tank (Figure 11–3)

The gas remaining in the tank experiences an adiabatic reversible expansion. Therefore, from Eq. 11–16

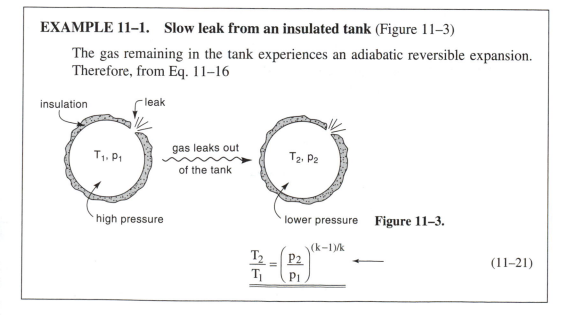

insulation   leak

T_1, p_1

gas leaks out
of the tank

T_2, p_2

high pressure   lower pressure   **Figure 11–3.**

$$\frac{T_2}{T_1} = \left(\frac{p_2}{p_1} \right)^{(k-1)/k} \longleftarrow \qquad (11\text{–}21)$$

EXAMPLE 11–2.   Rupture of a diaphragm in an insulated tank (Figure 11–4)

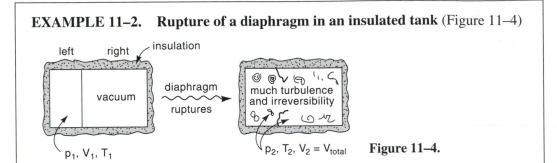

Figure 11–4.

If we focus on the left-hand side of the tank, we find that none of the equations we have developed apply and we are stuck; the same with the right-hand side. So let us look at the tank as a whole and see what we can say. This yields something useful. From the first law, Eq. 11–10, with $\Delta E_p = 0$ and $\Delta E_k = 0$, we get

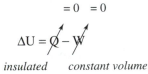

$$\Delta U = \cancel{Q} - \cancel{W}$$

insulated   constant volume

so from Eq. 11–10

$$T_2 = T_1 \text{ and } \frac{p_2}{p_1} = \frac{V_1}{V_2} \longleftarrow \tag{11–22}$$

EXAMPLE 11–3.   Slow leak between sections of an insulated tank
(Figure 11–5)

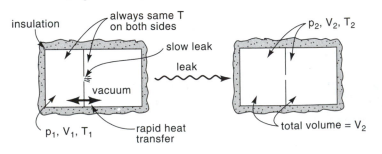

Figure 11–5.

For the whole tank, the first law, Eq. 11–10, with $\Delta E_p = 0$ and $\Delta E_k = 0$, gives

$$\Delta U = \cancel{Q} - \cancel{W}$$
$$\quad\;\underset{insulated}{}\quad\underset{constant\ volume}{}$$

with the $=0$ marks above Q and W.

and for an ideal gas with $\Delta U = 0$, Eq. 11–2 or 11–11 gives

$$T_2 = T_1 \quad \text{and} \quad \frac{p_2}{p_1} = \frac{V_1}{V_2} \quad\longleftarrow \tag{11–23}$$

EXAMPLE 11–4. **Heat involved in the slow isothermal reversible expansion of gas in a cylinder** (Figure 11–6)

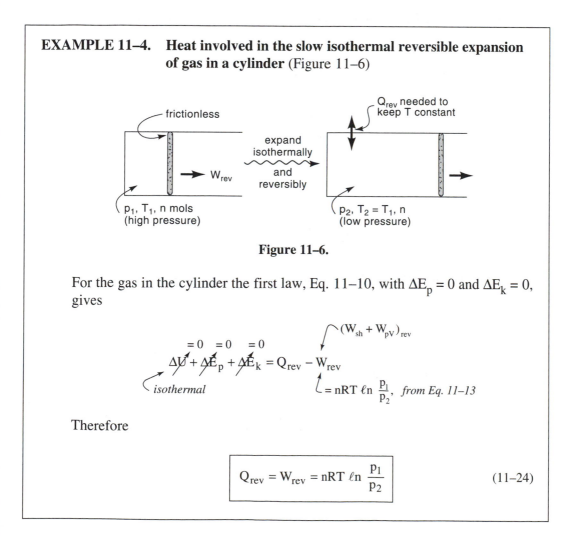

Figure 11–6.

For the gas in the cylinder the first law, Eq. 11–10, with $\Delta E_p = 0$ and $\Delta E_k = 0$, gives

$$\cancel{\Delta U} + \cancel{\Delta E_p} + \cancel{\Delta E_k} = Q_{rev} - W_{rev}$$

with $=0$ marks, ΔU isothermal, and $W_{rev} = (W_{sh} + W_{pV})_{rev} = nRT \ln \frac{p_1}{p_2}, \quad \text{from Eq. 11–13}$

Therefore

$$\boxed{Q_{rev} = W_{rev} = nRT \ln \frac{p_1}{p_2}} \tag{11–24}$$

EXAMPLE 11–5.   Leak between two interconnected insulated tanks
(Figure 11–7)

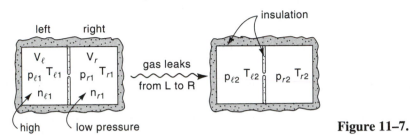

Figure 11–7.

Find how T_ℓ, T_r, p_ℓ, and p_r change.

Solution

From the ideal gas law and material balance

$$\left. \begin{array}{l} n_{\ell 1} = \left(\dfrac{pV}{RT}\right)_{\ell 1} \\[3mm] n_{r1} = \left(\dfrac{pV}{RT}\right)_{r1} \end{array} \right\} n_{\ell 1} + n_{r1} = n_{\text{total}} \qquad (11\text{–}25)$$

$$\left. \begin{array}{l} n_{\ell 2} = \left(\dfrac{pV}{RT}\right)_{\ell 2} \\[3mm] n_{r2} = \left(\dfrac{pV}{RT}\right)_{r2} \end{array} \right\} n_{\ell 2} + n_{r2} = n_{\text{total}} \qquad (11\text{–}26)$$

The gas remaining on the left-hand side expands adiabatically and reversibly, so

$$\frac{T_{\ell 2}}{T_{\ell 1}} = \left(\frac{p_{\ell 2}}{p_{\ell 1}}\right)^{(k-1)/k} \qquad (11\text{–}27)$$

Now apply the first law for the whole container. Because the container is isolated Eq. 3–1 gives

$$\sum E_2 = \sum E_1$$

or

$$\left(n_\ell\, u_\ell\right)_2 + \left(n_r\, u_r\right)_2 = \left(n_\ell\, u_\ell\right)_1 + \left(n_r\, u_r\right)_1$$

$$c_v\, T_1$$

and for an ideal gas Eq. 11–7 gives

$$\left(n_\ell \, T_\ell\right)_2 + \left(n_r \, T_r\right)_2 = \left(n_\ell \, T_\ell\right)_1 + \left(n_r \, T_r\right)_1 \qquad (11\text{–}28)$$

Given initial conditions, and in addition the final pressure on any one side, say $p_{\ell 2}$, the above equations will let you evaluate all else, $T_{\ell 2}$, p_{r2}, and T_{r2}.

All sorts of more complicated variations of Example 11–5 can and have been thought up by sadistic professors to torture their puzzling students. These problems have little practicality, but are used to test whether students can apply the equations in this chapter.

In all problems involving heat, work, or temperature change of an ideal gas during a process:

- Decide what is your system.
- Write the material balance expression, if needed.
- Write the pVT expression for ideal gases.
- Write the first law energy expressions for the ideal gas.
- If there is expansion or compression, do these occur adiabatically, isothermally, or reversibly? If so, write the corresponding expressions.

Then solve your problem. The above example is a typical example of this general procedure.

EXAMPLE 11–6.   Total work done by an expanding gas

A 2 liter plastic pop bottle contains air at 300 K and 11.5 bar gauge pressure. How much work could be done by this gas if you could expand it down to 1 bar

  (a) isothermally and reversibly

  (b) adiabatically and reversibly?

Solution

First let us determine the number of moles of air that we are dealing with:

$$n = \frac{pV}{RT} = \frac{(12.5 \times 10^5)(0.002)}{(8.314)(300)} = 1.00 \text{ mol}$$

We are now ready to determine the work done by the expanding air.

(a) *Isothermal expansion.* From Eq. 11–13 we get

$$W_{rev} = n\,RT\,\ell n\frac{p_1}{p_2}$$

$$= (1)\,(8.314)\,(300)\,\ell n\frac{12.5}{1} = \underline{6300\ J} \quad \longleftarrow$$

(b) *Adiabatic expansion.* From Eq. 11–17 we get

$$W_{rev} = \frac{n\,RT}{k-1}\left[1-\left(\frac{p_2}{p_1}\right)^{(k-1)/k}\right]$$

$$= \frac{(1)(8.314)(300)}{1.4-1}\left[1-\left(\frac{1}{12.5}\right)^{0.4/1.4}\right] = \underline{\underline{3205\ J}} \quad \longleftarrow$$

Note that this expanded gas is cooled. Also, note that isothermal expansion produces more work than does adiabatic expansion.

EXAMPLE 11–7.   Net work done by an expanding gas

The previous example calculated the work done by an expanding gas. However, in doing so the gas had to push back the 1 bar atmosphere. Let us now account for this work, subtract it from the work done, and thereby evaluate the useful work (shaft work) that could be extracted by this

(a) isothermal expansion.
(b) adiabatic expansion.

Solution

(a) *Isothermal expansion.* The work needed to push back the atmosphere is

$$W_{pV} = p_0(V_2 - V_1)$$

$$= (1\times10^5)\,(12.5\times0.002 - 0.002) = 2300\ J$$

So the reversible shaft work that can be extracted is

$$W_{sh} = 6300 - 2300 = \underline{4000\ J} \quad \longleftarrow$$

from Example 6

(b) *Adiabatic expansion*. First of all note that the final temperature of the expanded air is not 300 K. Equation 11–16 shows that it is

$$T_2 = T_1\left(\frac{p_2}{p_1}\right)^{(k-1)/k} = 300\left(\frac{1}{12.5}\right)^{0.4/1.4} = 146 \text{ K}$$

The work needed to push the surroundings back is

$$W_{pV} = p_0(V_2 - V_1)$$

$$= \left(1\times10^5\right)\left[12.5\left(\frac{146}{300}\right)0.002 - 0.002\right] = 1015 \text{ J}$$

So the shaft work when the gas expands is

$$W_{sh} = 3205 - 1015 = \underline{\underline{2190 \text{ J}}} \longleftarrow$$

from Example 6

EXAMPLE 11–8.   Explosions: The popping pop bottle

Find the energy release when a 2 liter plastic pop bottle pressurized to 11.5 bar gauge and 300 K explodes.

Solution

Referring back to Example 11–6, we first see that the bottle contains one mole of air. Then, in looking at the energy change we must first decide what assumptions to make about the explosion. Is it reversible or not; is it isothermal or adiabatic?

We are tempted to assume reversible operations as was done in Examples 11–6 and 11–7 because then we can use the equations in this chapter. However, that would be silly, because if anything is irreversible it's an explosion. As for the choice between adiabatic or isothermal, the explosion is so rapid that it certainly is more reasonable to assume an adiabatic process. So we assume that the explosion is best approximated by a highly irreversible adiabatic process. Then from first law

$$\overset{= 0}{\underset{}{\Delta U = \cancel{Q} - W_1}} \quad \text{total work}$$

$$(11\text{–}29)$$

All this work is used to push back the atmosphere. No useful or shaft work is done since this is a highly irreversible process, so

$$W_2 = \int p \, dV = p_{surr}(\Delta V) \tag{11-30}$$

Now the final temperature is the unknown in these two expressions. So inserting numerical values gives

$$W_1 = -\Delta U = n \, c_v (T_{initial} - T_{final})$$
$$= (1) \, (29.099 - 8.314) \, (300 - T_{final}) \tag{11-31}$$

and

$$W_2 = p_{surr}(V_{final} - V_{initial})$$
$$= 10^5 \left[12.5 \times 0.002 \left(\frac{T_{final}}{300} \right) - 0.002 \right] \tag{11-32}$$

Solving Eqs. 11-31 and 11-32 simultaneously, noting that $W_1 = W_2$, gives the final temperature to be

$$T = 221 \text{ K}$$

and the work done, from Eq. 11-31, is

$$\underline{W = 1642 \text{ J}} \longleftarrow$$

We included some hidden assumptions and simplifications in our analysis.

- That the gas is ideal at this high pressure—reasonable
- That the work needed to tear, heat, and even melt some of the plastic is small—probably reasonable
- The frictional loss and energy in the shock wave (and sound) is negligible—questionable.

These ignored factors should not seriously affect our answer.

Of course, if you had been holding the plastic bottle when it exploded, it would have done some "useful" work in blowing off your hand. In that case the work done by the gas would have been somewhere between the irreversible and reversible values of 1642 J (Example 11-8) and 3205 J (Example 11-6).

PROBLEMS

1. Which requires more heat input, to heat a gas in a constant pressure cylinder or in a constant volume bomb?

2. Googliodifaker ($\overline{mw}$ = 124 gm/mol) is an ideal gas with c_p = 309 J/kg·K. I need to find the ratio of specific heats for this gas, or k = c_p/c_v. Would you please calculate it for me?

3. At 200°C propene (C_3H_4) has a specific heat

$$c_p = 1940 \text{ J/kg·K.}$$

Find k, the ratio of specific heats of propene at these conditions.

4. A frictionless piston-cylinder device contains 1 m³ of a bimolecular gas (c_p = 28.314 J/mol·K) at p = 99 768 Pa and 27°C. Heat is added to the gas, which causes its pressure to rise. Then the gas is allowed to expand, and heat is added or removed, as shown in the sketch below. How much heat was added during this whole operation?

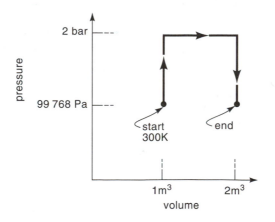

5. A cylinder with piston contains 2 liters of nitrogen gas (c_p = 1039 J/kg·K) at 3 atm. The gas is to be heated at constant pressure from 20°C to 199°C. How much heat is added to the gas in the cylinder?

6. A large insulated tank is divided into two sections by a diaphragm. Section A of 60 liters contains methane gas, CH_4, at 400 kPa and 800 K. The other section, B, of 180 liters is completely evacuated. The diaphragm bursts and the CH_4 is redistributed. What is the final T and p in the tank?

7. A diaphragm divides a rigid, well insulated 2 m³ tank into two equal parts. The left side contains an ideal gas (c_p = 30 J/mol·K) at 10 bar and 300 K.

The right side contains nothing—it's a vacuum. A small hole forms in the diaphragm, gas slowly leaks out from the left side, and eventually the temperature in the whole tank equalizes. What is this final temperature?

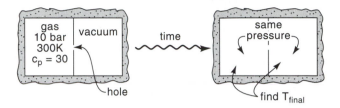

8. An insulated tank (166.28 liter) is divided by a diaphragm into two equal parts. One contains H_2 at 900 kPa and 300 K; the other contains H_2 at 400 kPa and 400 K. The diaphragm bursts and the gases mix. What is the final pressure in the system?

9. A high-powered bullet cleanly punctures a *well* insulated gas storage tank containing gas (c_p = 31.622 J/mol·K) at 10 atm and 300°K. Gas leaks from this tank and the hole is finally stoppered when the tank is at a pressure of 5 atm. What is the temperature of the gas remaining in the tank?

10. An ideal gas (c_p = 30 J/mol·K, 24.6 K, 0.9986 bar) in a cylinder is compressed slowly without friction, but isothermally from 3.33 m^3 down to 1 m^3. Find the work required to do this.

11. Six liters of a bimolecular ideal gas at 1 bar (c_p = 29.1 J/mol·K) are compressed in a cylinder until the volume becomes 2 liters. A pressure gauge in the cylinder shows that

$$pV^2 = \text{constant}$$

during the compression. Find the work done by the gas during this compression.

12. A hot ideal gas in a cylinder is cooled and compressed by pushing in a piston. During this operation the following data regarding the conditions in the cylinder are recorded

T, °C	V, liters	p, atm
303	24	12
189	21	11
123	18	11
63	14	12
63	12	14

What is the heat interchange of gas with its container and surroundings?

Data For the gas: $c_p = 26.06$ J/mol·K, $\overline{mw} = 0.029$ kg/mol.

13. 1 mole of nitrogen in a piston-cylinder assembly is compressed—not adiabatically, not isothermally, but somewhere in between—according to the relationship

$$pV^{1.2} = \text{constant}$$

from 100 kPa and 300 K to 1 MPa.

Find the heat involved in this process if there is any. If there is some, is it removed or added?

14. Imagine a rigid tube 0.76 m high filled with mercury ($\rho = 13\ 600$ kg/m^3, considered to be incompressible) with a tiny bubble of ideal gas attached to the bottom of the tube by surface tension. The pressure at the bottom of the tube is 3 atm, at the top 2 atm.

I tap the tube, the tiny bubble of ideal gas is freed from the surface and rises to the top of the tube. What is the pressure at the bottom of the tube after this happens? Assume that the temperature is unchanged.

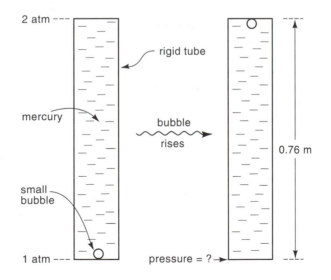

15. Let us extend the previous problem to a more interesting situation. Suppose we had a 1.52 m long rigid tube filled with incompressible mercury except for a small 1 atm bubble of air at the top.

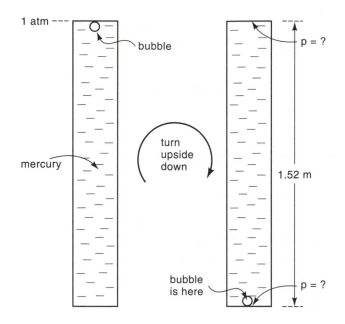

Now carefully turn the tube upside down so that the bubble remains unmoved in the tube (now at the bottom). What is the situation at the two ends of the tube—the pressure and the bubble?

16. *Airbags for automobiles.* In a head-on auto collision, the front of the car is pushed back. This activates a primer that triggers the rapid decomposition (20 ~ 30 ms) of an exotic chemical, sodium azide, which releases gas to fill the air bag

$$2\,NaN_3(s) \xrightarrow{\text{primer}} 3N_2 + \text{etc.}$$

This is a very delicate and complicated process that is widely used today.

As an alternative, let us look into the possibility of using a compressed gas to inflate the air bag. For this suppose a cylinder contains a highly compressed ideal gas, air, at 50 MPa and 20°C. In an automobile collision the cylinder is pierced and gas flows into the bag. If the initial volume of air in the bag is zero and if it blows up to 60 liters at about 1 bar, what should be the volume of the high pressure cylinder and what will be the temperature of the air bag air when inflated?

Note: Air bags have holes in their sides to let air escape, otherwise a person would bounce into it and snap back hard, which would be dangerous and even lethal. But when first filled the bag inflates without any appreciable escape of air. Also, doesn't this process seem simpler and more reliable than the chemical reaction process being used at present?

17. The sketches below show two ideal experiments, both done in perfectly insulated containers. Do you think that $T_2' > T_2$, or $T_2' < T_2$, or $T_2' = T_2$? Discuss how you came to this conclusion.

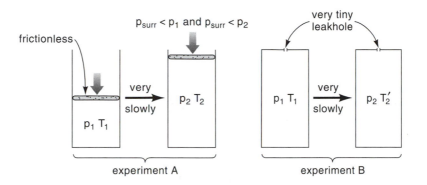

CHAPTER 12
ENGINEERING FLUIDS

This chapter considers a variety of topics, all related to the properties of gases and liquids, that are useful for engineering purposes. We take up in turn

- Mixture of ideal gases
- Combination of solids, liquids, and vapors
- Two important engineering fluids
- Nonideal gases

We start with the pure material ideal gas that well represents all gases at not too high pressure. As shown in Chapters 2 and 11, the equation that represents this is simply

$$pV = nRT \quad \text{or} \quad p\,\overline{mw} = \rho RT \tag{12–1}$$

Let us now look at mixtures of ideal gases.

A. MIXTURE OF IDEAL GASES

With gases it is much more convenient to deal with moles not masses, thus mole fractions, volume fractions, and partial pressures. So consider a mixture of ideal gases consisting of n_A moles of A, n_B moles of B, and so on in a box at T K and π bar.

The *partial pressure of A*, p_A, is defined as the pressure exerted by A if it were alone in the box and
The *pure component volume*, V_A, is defined as the volume of the container that contains just A alone at T and π.

Then, for mixtures of ideal gases we have

Dalton's law: the total pressure, $\pi = p_A + p_B + \ldots$ [Pa = N/m^2]   (12–2)

Amagat's law: the total volume, $V = V_A + V_B + \ldots$ [m^3]   (12–3)

As an illustration, consider a mixture of 80 mol% of A and 20 mol% of B in a 10 lit box at 5 bar. Figure 12–1 illustrates what these two laws mean.

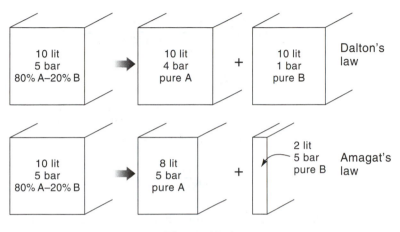

Figure 12–1.

Also the mole fraction of A is

$$\left(\begin{matrix} \text{mole fraction} \\ \text{of A} \end{matrix}\right) = \frac{p_A}{\pi} = \frac{V_A}{V} \qquad (12–4)$$

So for ideal gas mixtures

- Partial pressures are additive
- Pure component volumes are additive
- Mole fractions = pressure fractions = pure component volume fractions

This means that each component of a gas mixture behaves independently as though it alone is present in the container. So the ideal gas law expression for a mixture can be written

$$(p_A + p_B + \ldots)\, V = (n_A + n_B + \ldots)\, RT \qquad (12–5)$$

EXAMPLE 12–1.　Density of a mixture, composition given in mass fractions

Find the mean molecular mass $\overline{mw}$ and density ρ at 200 kPa and 88°C of a 4 wt% of hydrogen and 96 wt% of oxygen mixture.

Solution

Since we are given masses, let us take as our basis of calculation 100 kg of mixture. Then

$$\text{moles of } H_2 = 4 \text{ kg} \left(\frac{1 \text{ mol}}{0.002 \text{ kg}} \right) = 2000 \text{ mol}$$

$$\text{moles of } O_2 = 96 \text{ kg} \left(\frac{1 \text{ mol}}{0.032 \text{ kg}} \right) = 3000 \text{ mol}$$

$$\overline{\text{Total} = 5000 \text{ mol}}$$

Therefore, the mean molecular mass of the mixture is

$$\overline{mw} = \frac{100 \text{ kg}}{5000 \text{ mol}} = 0.020 \frac{\text{kg}}{\text{mol}} \quad \longleftarrow$$

To find the density we need the mass and volume. The mass is 100 kg, the volume from the ideal gas law is

$$V = \frac{n_{total}RT}{p} = \frac{(5000)(8.314)(361)}{200\,000} = 75 \text{ m}^3$$

Therefore

$$\rho = \frac{m}{V} = \frac{100 \text{ kg}}{75 \text{ m}^3} = 1.33 \frac{\text{kg}}{\text{m}^3} \quad \longleftarrow$$

EXAMPLE 12–2.　Density of a mixture, composition given in mol fractions

Find the mean molecular mass and density at 131°C and 1.2 bar of a 60 mol% hydrogen, 40 mol% oxygen mixture.

Solution

Since we are given molar units, let us take as our basis 100 moles of mixture. Then

$$\text{mass of } H_2 = (60 \text{ mol}) \left(\frac{0.002 \text{ kg}}{\text{mol}} \right) = 0.12 \text{ kg}$$

$$\text{mass of } O_2 = (40 \text{ mol}) \left(\frac{0.032 \text{ kg}}{\text{mol}} \right) = 1.28 \text{ kg}$$

$$\overline{\text{Total mass } = \ 1.40 \text{ kg}}$$

Therefore

$$\overline{mw} = \frac{\text{total mass}}{\text{total mol}} = \frac{1.4}{100} = 0.014 \ \frac{\text{kg}}{\text{mol}} \quad \longleftarrow$$

To evaluate the density we need the volume of the mixture. From the ideal gas law

$$V = \frac{n \, R \, T}{p} = \frac{100 \, (8.314) \, (404)}{120 \, 000} = 2.8 \text{ m}^3$$

So the density

$$\rho = \frac{m}{V} = \frac{1.40}{2.8} = 0.5 \ \frac{\text{kg}}{\text{m}^3} \quad \longleftarrow$$

B. PURE MATERIAL GOING FROM SOLID TO LIQUID TO GAS

1. The p-T Diagram

For pure materials that do not react or decompose, Figure 12–2 shows what phase changes take place when p or T changes. Most materials follow the behavior of Figure 12–2a but a very important few, such as water and vinyl plastic (PVC), follow Figure 12–2b.

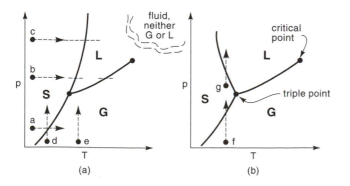

Figure 12–2. p-T diagram for (a) a normal substance that shrinks on freezing; (b) for an unusual substance that expands on freezing.

These sketches show two unique conditions, the triple point and the critical point. The triple point is that one particular condition where gas, liquid, and solid can all coexist. The critical point is that particular condition above which gas and liquid merge into one phase that we call fluid, or supercritical fluid.

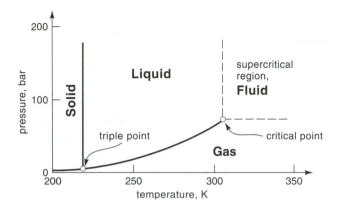

Figure 12–3. p-T diagram for CO_2, to scale.

The sketches of Figure 12–2 are distorted and not drawn to scale, for the purpose of illustrating the special features, such as slopes of the curves. In reality, the figures are more like those shown in Figures 12–3 and 12–4.

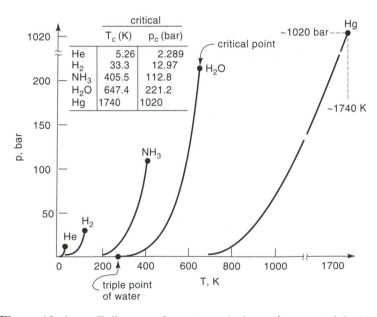

	critical	
	T_c (K)	p_c (bar)
He	5.26	2.289
H_2	33.3	12.97
NH_3	405.5	112.8
H_2O	647.4	221.2
Hg	1740	1020

Figure 12–4. p-T diagrams drawn to scale for various materials. Note that the G-L region dominates, and the region dealing with triple points and solids is not even seen.

Since we are most familiar with water, let us use it to illustrate the ideas being developed here. Thus, the triple and critical points for water have been measured to be

$$
\text{Triple point}
\begin{cases}
p = 0.610 \text{ Pa} \\
T = 0.01°C \\
\rho_\ell = 1000.0 \text{ kg/m}^3 \\
\rho_s = 916.8 \text{ kg/m}^3 \\
\rho_g = 0.00485 \text{ kg/m}^3
\end{cases}
\qquad
\text{Critical point}
\begin{cases}
p = 221.3 \text{ bar} \\
T = 647.3 \text{ K} \\
\rho = 323 \text{ kg/m}^3
\end{cases}
$$

Heating a pure material at constant pressure. If we start with very cold solid at low pressure (point a in Figure 12–2a) and heat it, we go directly from solid to gas. At higher pressure, between triple and critical (point b in Figure 12–2a), a rise in temperature will melt the solid to give liquid and then gas. Finally, at very high initial pressure, above the critical (point c in Figure 12–2a), the solid transforms directly to a fluid.

Increase in pressure on a pure material at constant temperature. For just about all materials at low temperature an increase in pressure transforms the gas directly into solid (point d in Figure 12–2a).

At a higher temperature an increase in pressure changes the gas to liquid and then to solid (point e in Figure 12–2a). However, most interestingly, water and PVC plastic are two of the very few exceptions to this behavior (Figure 12–2b). So at intermediate temperature a rise in pressure will transform the gas to solid and then to liquid (points f and g in Figure 12–2b), not the other way around. Thus, liquid water is denser than ice!![1]

---

[1] If it were not for this amazing property, icebergs would not float, any ice formed in winter would sink to the bottom of the oceans, ice would then accumulate from the bottom up, and there would only be a thin liquid layer at ocean surfaces. What this means is that only the most elementary forms of life could have formed on earth. So the fact that you are alive and reading this page is due to the fact that water follows Figure 12–2b, not 12–2a. What good fortune.

Also, no ice skating would be possible. Think about it. An ice skate puts a high pressure on a narrow groove of ice and melts it, all at below 0°C (point g of Figure 12–2b). After the skate passes by the pressure returns to normal and the high pressure water formed returns to low pressure ice.

As another example of this phenomenon, when old records are played, needles scratch the record, creating fuzz and worn-out needles and records. However, today's vinyl (PVC) records melt under pressure of the diamond needle that, though only weighing 2 gm, exerts this force on a very small area, giving a pressure of about 600 bar on the vinyl. Solid vinyl melts under this pressure and immediately refreezes as the needle passes by. Hence no fuzz and no worn-out records.

2.   The p-V Diagram

In an alternative representation of the properties of pure materials, Figure 12–5 (not to scale) shows how the volume of material changes for the two types of materials, those that shrink on freezing (Figure 12–5a) and those that expand on freezing (Figure 12–5b).

This graph, in particular the liquid-vapor region, is most useful in thermo for the evaluation of the performance of engines. We will meet it again.

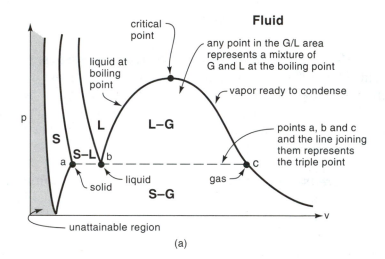

(a)

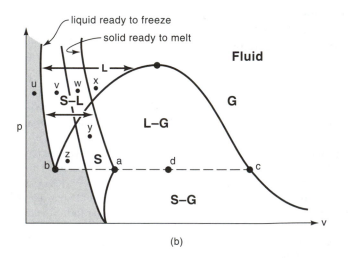

(b)

Figure 12–5.   The p-v diagram (a) for a normal pure substance, one that shrinks on freezing, and (b) for a substance that expands on freezing. These sketches are not to scale.

C. TWO IMPORTANT ENGINEERING FLUIDS—WATER AND HFC-134a

By compressing and vaporizing a liquid, and then expanding and condensing a vapor, engineers discovered how to convert heat to work. Think of the steam engine. Opposed to this, by doing work on a fluid, other engineers discovered how to pump heat "uphill" from a lower to a higher temperature. Think of an air conditioner or a refrigerator where heat is pumped from the refrigerator box to the outside.

For getting work from heat (burning of wood, coal, oil) we needed a fluid that would vaporize and condense at above room temperature, say up to 300°C. Water is ideal for this purpose.

For refrigeration we needed a fluid that would vaporize and condense somewhere below room temperature, say down to −40°C. Ammonia has this property and was used practically exclusively long ago. However, it is poisonous and corrosive. So DuPont researchers came up with a family of fluids that they called Freons (note the capital letter in the name). They are hydrocarbons with attached fluorine and chlorine atoms, hence their chemical family name was chlorofluorocarbons, or CFCs. The most useful of these was Freon-12, with chemical formula

$$
\begin{array}{c}
\text{Cl} \\
| \\
\text{Cl} - \text{C} - \text{F} \ \ldots \ \text{dichlorodifluoromethane} \\
| \\
\text{F}
\end{array}
$$

They came up with many other Freons, such as Freon-113 and Freon-23.

Freons were hailed as ideal refrigerants since they were nonpoisonous, noncorrosive, and chemically inert. They replaced ammonia in most applications and have been used for decades without any problems. Recently, however, it was discovered that when old refrigerators and home and auto air conditioners were junked, their spilled Freons vaporized, rose in the atmosphere, and their chlorine attacked the ozone layer that was protecting the earth against deadly solar radiation. So Freons were found to be harmful, had to be banned and replaced, but by what?

DuPont engineers again performed their chemical magic and came up with a new family of refrigerants, the hydrofluorocarbons, or HFCs, which DuPont called Suvas. These compounds were hydrocarbons with attached fluorine atoms, but with no chlorine. The most useful of these is HFC-134a, or Suva-134a, which has the chemical formula

```
   F     F           ⌐ attached to the first carbon
   |     |          ╱
   |     |         ↓
F — C — C — H  ... 1,1,1,2 tetrafluoroethane, $\overline{mw}$ = 0.102 kg/mol
   |     |         ↑
   |     |          ╲
   F     H           ⌐ attached to the second
```

Chemical plants built to produce Freons are all closing down and new plants are being built in Europe and the United States to produce Suvas for worldwide use.

To determine what can be gotten from engineering fluids—the work obtainable, amount of cooling, and so on—engineers had to know the properties of their working fluids, properties such as density and enthalpy at given pressure and temperature. These are tabulated at the back of most thermo texts.

- Prior to 1930 the tabulations were for NH_3 and water.
- Between 1930 and 1990 the tabulations were for NH_3, Freon-12, and water.
- In this book we give it for HFC-134a and water, since Freons represent past history.

The examples and problems that follow give practice in the use of these tables. Chapter 17 extends this practice to the use of the entropy function, which is introduced in Chapter 15.

D. MIXTURES OF PHASES AND THERMO TABLES

First of all, let us mention that all values in the thermo tables are tabulated per kg of pure materials. These are called *specific properties*. Thus, for a pure gas, some prefer to call these vapors, and for a pure liquid

$$v_g = \frac{m^3 \text{ gas}}{\text{kg gas}} \quad \text{and} \quad v_\ell = \frac{m^3 \text{ liquid}}{\text{kg liquid}}$$

Next, note that two phases of a pure substance can only exist together at the boundary separating phases in Figure 12–5. These are called *saturation conditions*.

Also, for a gas-liquid mixture the mass fraction of gas x_g is called the *quality* of the mixture. Thus

$$\text{quality} = x_g = \left(\begin{array}{c} \text{mass fraction} \\ \text{of gas in the} \\ \text{mixture} \end{array} \right) \left[\frac{\text{kg gas}}{\text{kg total}} \right]$$

Similarly,

$$x_\ell = \begin{pmatrix} \text{mass fraction} \\ \text{of liquid in} \\ \text{the mixture} \end{pmatrix} = 1 - x_g \quad \left[\frac{\text{kg liquid}}{\text{kg total}} \right]$$

We are now ready to calculate desired quantities of mixtures from the thermo tables in the Appendix. Thus, for a mass m_{total} of a mixture having m_g kg of gas and m_ℓ kg of liquid, and having a volume V_{total} consisting V_g m³ of pure gas and V_ℓ m³ of pure liquid, we can write the following relationships

$$m_{total} = m_g + m_\ell \quad \text{[kg of mixture]} \tag{12-6}$$

$$V_{total} = V_g + V_\ell = v\, m_{total} \quad \left[m^3 \text{ total} \right] \tag{12-7}$$

$$v = x_g v_g + x_\ell v_\ell = \frac{V_{total}}{m_{total}} \quad \left[\frac{m^3 \text{ total}}{\text{kg of mixture}} \right] \tag{12-8}$$

or in slightly different form

$$x_g = \frac{v - v_\ell}{v_g - v_\ell} \quad \text{and} \quad x_\ell = \frac{v_g - v}{v_g - v_\ell} \quad [-] \tag{12-9}$$

The volume of gas and of liquid in the mixture is then

$$V_g = x_g v_g m_{total} \quad \text{and} \quad V_\ell = x_\ell v_\ell m_{total} \quad \left[m^3 \text{ of g or } \ell \right] \tag{12-10}$$

Other properties of phase mixtures, besides volumes, can be obtained from pure substance properties with equations similar to those above.

EXAMPLE 12–3. Gas-liquid mixture in a container

A 20-liter flask contains no air, nothing but 2 kg of pure water, and it is at 245°C. How much liquid water does it contain?

Solution

Given $V_{total} = 0.02$ m³ and $v = 0.02/2 = 0.01$ m³/kg, then at 245°C, the table in the appendix gives

$$v_g = 0.050\ 13\ \text{m}^3/\text{kg gas}$$

$$v_\ell = 0.001\ 25\ \text{m}^3/\text{kg liquid}$$

$$p = 39.73\ \text{bar}$$

Then, from Eq. 12–9

$$x_\ell = \frac{v_g - v}{v_g - v_\ell} = \frac{0.050\ 13 - 0.010}{0.050\ 13 - 0.001\ 25} = 0.8210 \text{ kg liquid/kg total}$$

Therefore, mass and volume of liquid are

$$m_\ell = x_\ell m_{total} = \quad (0.8210)\,(2) \quad = 1.6420 \text{ kg liquid} \quad\longleftarrow$$

$$V_\ell = m_\ell v_\ell \quad = (1.642)\,(0.001\ 25) = 0.0021 \text{ m}^3 \text{ liquid} \longleftarrow$$

or 10.5 vol% liquid $\longleftarrow$

Note. We represent this answer as point a on Figure 12–5, or in Figure 12–6, and this represents a mixture of liquid (point b) and gas (point c).

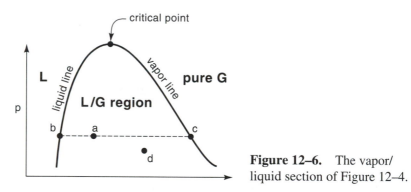

Figure 12–6.   The vapor/liquid section of Figure 12–4.

In general, whenever one has a mixture, such as point a, inside any two-phase region G/L, G/S, or L/S, it represents a mixture of the left and the right extreme of the region.

EXAMPLE 12–4.   Evaluation of work done by an expanding fluid

A red painted cylinder with piston contains 3 kg of water at 20°C and 10 bar (Figure 12–7). Heat is added, the water heats up, boils, becomes steam, and the piston moves out until the temperature reaches 300°C, all this at 10 bar. For the 3 kg of fluid (ignore any heat added to the metal of the piston and cylinder)

(a) Find ΔU.

(b) How much work is done by the fluid?

(c) How much heat is added to the fluid?

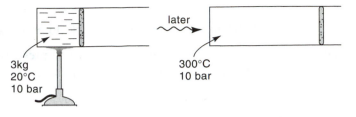

Figure 12–7.

Solution

From the steam tables, at 10 bar

not given in tables   *reasonable assumption for liquids*

$$v_{20°C,\ 10\ bar} \cong v_{20°C,\ 2.3\ kPa} = 0.001\ 002\ m^3/kg$$

$$v_{300°C} = 0.2579\ m^3/kg$$

$$h_{20°C} \cong 83.96\ kJ/kg$$

$$h_{300°C} = 3051.2\ kJ/kg$$

So the work of expansion done by the fluid is

$$W = \int p\,dV = p(V_2 - V_1) = mp(v_2 - v_1)$$

$$= 3\ kg\ (10^6\ Pa)\ (0.2579 - 0.0010)\ m^3/kg = \underline{770\ 700\ J} \quad \longleftarrow \quad (b)$$

The change in internal energy of the fluid

$$\Delta U = m(\Delta h - \Delta pv)$$

$$= 3kg\left[(3051.2 - 83.96)\ 10^3\ J/kg - 10^6\ Pa(0.2579 - 0.0010)\ m^3/kg\right]$$

$$= \underline{8131\ 020\ J} = \underline{8131\ kJ} \quad \longleftarrow \quad (a)$$

From the first law for batch systems we have in general

$$\Delta U + \cancel{\Delta E_p} + \cancel{\Delta E_k} = Q - W \tag{7–5}$$

and at constant pressure

$$Q = \Delta U + W = \Delta H = m(h_2 - h_1) \tag{7–7}$$

Replacing values gives

$$Q = m\Delta h = 3\ kg\ (3051.2 - 83.96)\ kJ/kg$$

$$= \underline{8901.72\ J} = \underline{8.9\ kJ} \quad \longleftarrow \quad (c)$$

EXAMPLE 12–5.   Failed attempt to evaluate the work done by expansion

Liquid HFC-134a at −10°C and 2.01 bar is heated and pressurized to 140°C and 10 bar. For each kilogram of this fluid enjoying this change

  (a) determine the increase of internal energy, Δu.
  (b) determine the heat added, q.
  (c) determine the work done on expansion, w.

Solution

From the tables we have

$$\text{at } -10°C \text{ and 2.01 bar} \begin{cases} v_1 = 0.000\ 75 \text{ m}^3\text{kg} \\ h_1 = 186.711 \text{ kJ/kg} \end{cases}$$

$$\text{at } 140°C \text{ and 10 bar} \begin{cases} v_2 = 0.031\ 55 \text{ m}^3/\text{kg} \\ h_2 = 525.6 \text{ kJ/kg} \end{cases}$$

So the change in internal energy is

$$\Delta u = \Delta h - \Delta pv = h_2 - h_1 - (p_2 v_2 - p_1 v_1)$$

$$= (525.6 - 186.711) - [(1000)\ (0.031\ 55) - 200(0.000\ 75)]$$

$$\underset{kJ/kg}{\uparrow} \qquad\qquad \underset{kPa}{\uparrow} \quad \underset{m^3/kg}{\uparrow}$$

$$= \underline{307.49 \text{ kJ/kg}} \quad \longleftarrow$$

To calculate q and/or w we write the first law

$$\Delta u + \Delta \cancel{e}_p + \Delta \cancel{e}_k = q - w$$

or

$$\Delta u = q - w$$

Now the work term is given as

$$w = \int_{0.000\ 75}^{0.031\ 55} p\ dv$$
$$\text{↖} \textit{changes from 2 to 10 bar}$$

However, p is not constant and can take any one of an infinite number of paths as the volume changes. This means that we cannot evaluate the work done. So Δu can be evaluated, w − q can be evaluated, but not q or w separately. This is illustrated in Figure 12–8.

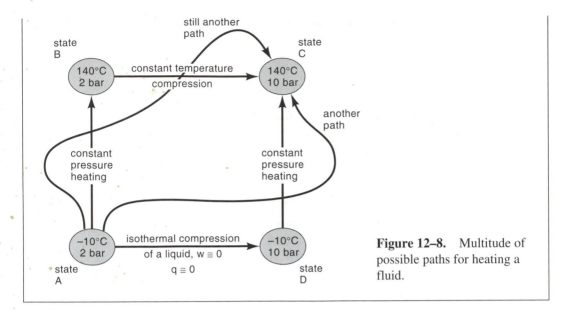

Figure 12–8.   Multitude of possible paths for heating a fluid.

Note the difference in the two previous examples. In the first the pressure is kept constant so the work can be evaluated; in the second example this is not so.

E.   EXTENSIONS TO OTHER ENGINEERING FLUIDS

With the use of thermodynamic tables we can relate the work done, heat involved, phase change, enthalpy change, and so on, with any change of pressure and temperature. We have shown how to do this with just two fluids, HFC-134a and water. For other useful engineering materials, such as NH_3 and other Suvas (HP62, HCFC-123, HFC-125), oxygen, nitrogen, and so on, just obtain their particular thermodynamic tables and use the same methods as outlined in this chapter.

It should be noted that thermodynamic tables are not usually prepared for very high pressures and very high temperatures, those beyond the critical point. At these extremes we must use the methods of the next section.

F.   HIGH PRESSURE AND NONIDEAL GAS BEHAVIOR

The ideal gas law, treated in Chapters 2 and 11, is very useful because it represents all gases at not too high a pressure. All you need to know is the molecular weight of the gas; the rest follows. This law thus represents a powerful generalization.

To account for the deviation from the ideal gas law at high pressures we introduce a correction factor to the ideal gas law. We call this the compressibility factor, as follows

$$pV = z\,nRT \quad \text{or} \quad pV(\overline{mw}) = z\,mRT$$

$$\uparrow$$
$$\text{compressibility factor}$$

$$(12\text{–}11)$$

At high pressure it was found that gases exhibited similar z values for similar fractional deviation of the gases from the critical point, or at similar p/p_c and T/T_c values. We call these ratios

$$\left.\begin{array}{l}\text{the reduced pressure}: \quad p_r = p/p_c \\[2mm] \text{the reduced temperature}: \quad T_r = T/T_c\end{array}\right\} \qquad (12\text{–}12)$$

Figure 12–9 gives experimental values of z at various T_r and p_r values. This chart shows that the maximum deviation of a factor of 4 or 5 from ideal gas behavior occurs at close to the critical point.

To calculate the property of any real gas you need to know its molecular weight and its critical properties. These are given in Table 12–1.

TABLE 12–1. Critical Properties of Some Common Gases

	Chemical Formula	Molecular Mass $\overline{mw}$, kg/mol	Critical Pressure p_c, bar	Critical Temperature T_c, K
Air	—	0.0289	37.7	132
Ammonia	NH_3	0.0170	113.0	406
Benzene	C_6H_6	0.0781	48.3	562
n-Butane	C_4H_{10}	0.0581	37.0	426
Carbon dioxide	CO_2	0.0440	74.0	304
Carbon monoxide	CO	0.0280	35.0	134
Chlorine	Cl_2	0.0709	77.1	417
Ethane	C_2H_6	0.0301	49.0	305
Ethyl alcohol	C_2H_5OH	0.0461	63.9	516
Helium	He	0.004	2.29	5.26
HFC-134a	CH_2FCF_3	0.102	40.6	374
Hydrogen	H_2	0.0020	13.0	33.3
Methane	CH_4	0.016	46.4	191
Methanol	CH_3OH	0.0321	79.7	513
Nitrogen	N_2	0.0280	33.9	126
n-Octane	C_8H_{18}	0.1142	24.9	569
Oxygen	O_2	0.0320	50.4	154
n-Pentane	C_5H_{12}	0.0722	33.4	470
Propane	C_3H_8	0.0441	42.6	370
Water	H_2O	0.0180	221.3	647

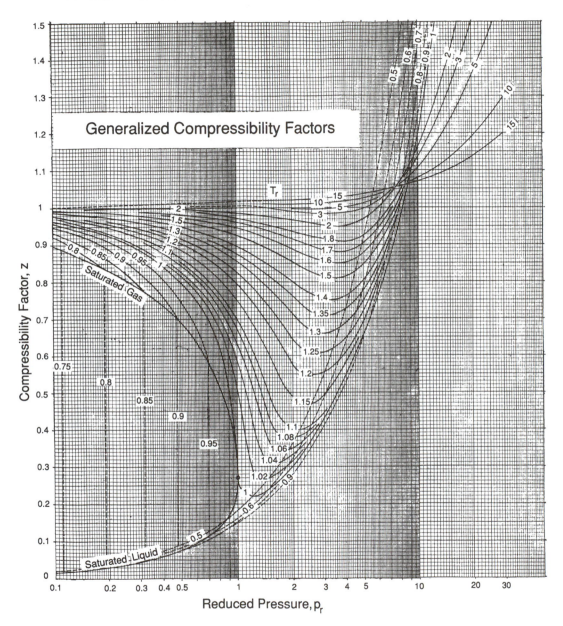

Figure 12–9.   Compressibility factor for gases at high pressure. Adapted from G. J. Van Wylen and R. E. Sonntag, *Fundamentals of Classical Thermodynamics, 2/e, S.I. Version.* John Wiley, 1976.

A much deeper and more detailed treatment of nonideal gas behavior, including mixtures of gases, is given by Hougen and Watson.[2]

The following examples show how to use the compressibility factor equation.

EXAMPLE 12–6.   Volume unknown, nonideal gas

Estimate the volume of 1 kg of carbon monoxide at 71 bar and 147.4 K.

Solution

Using Table 12–1 we find

$$p = 71 \text{ bar} \quad p_c = 35.4 \text{ bar} \quad p_r = \frac{71}{35.4} = 2.0$$

$$T = 147 \text{ K} \quad T_c = 134 \text{ K} \quad T_r = \frac{147}{134} = 1.10$$

From Figure 12–9 we find the compressibility factor

$$z = 0.4$$

So from Eq. 12–11

$$V = \frac{m \, z \, RT}{mw \, p}$$

$$= \frac{(1) \, (0.4) \, (8.314) \, (147)}{(0.028)\left(71 \times 10^5\right)} = \underline{\underline{0.002 \ 46 \text{ m}^3}} \quad \longleftarrow$$

EXAMPLE 12–7.   Pressure unknown, nonideal gas

Find the pressure needed to compress 300 liters of 7°C and one bar air to 1 liter at −115°C.

Solution

The original air (state 1) is at low enough pressure to be ideal, so take the compressibility factor $z_1 = 1$. However, at high pressure (state 2) we suspect nonideal behavior, so using Table 12–1 we find

---

[2]O. A. Hougen and K. M. Watson, *Chemical Process Principles, Part II* (New York: Wiley, 1947).

$$p_{r,2} = \frac{p_2}{37.7} \quad \text{with } p_2 \text{ unknown}$$

$$T_{r,2} = \frac{158}{132} = 1.20$$

Next, from Eq. 12–11

$$\frac{p_2 V_2}{p_1 V_1} = \frac{z_2 n \, RT_2}{z_1 n \, RT_1}$$

or

$$\frac{p_2(0.001)}{10^5(0.3)} = \frac{z_2 158}{1(280)} = 16\ 928\ 571\ z_2$$

or in reduced form

$$p_{r,2} = 4.55\ z_2 \qquad\qquad (12\text{–}13)$$

Now solve Eq. 12–13 by trial and error.

Given $T_{r,2}$	Guess $p_{r,2}$	From Figure 12–8 z_2	$\dfrac{p_{r,2}}{z_2}$
1.2	2.0	0.57	$\dfrac{2.0}{0.57} = 3.5$
1.2	1.0	0.80	1.25
1.2	2.5	0.55	4.55 ⟵ *correct value*

Therefore $p_2 = p_{r,2}\, p_c = 2.5\,(37.7) = \underline{94.25 \text{ bar.}}$ ⟵

From ideal gas law you would find 169 bar, which is quite different.

EXAMPLE 12–8.   Temperature unknown, nonideal gas

560 gm of dry ice (solid CO_2) are introduced into an evacuated 2-liter container. The temperature rises, and CO_2 vaporizes. If the pressure in the tank is not to exceed 111 bar, what should be the maximum allowable temperature of the tank?

Solution

At this high pressure we certainly expect to have nonideal gas behavior, so with values from Table 12–1 and the data given

$$p = 111 \text{ bar} \quad p_c = 74 \text{ bar} \quad p_r = \frac{111}{74} = 1.5$$

$$T = ? \quad\quad T_c = 304 \text{ K} \quad T_r = \frac{T}{304}$$

At this point we cannot evaluate z. So let's try a flanking maneuver by turning to Eq. 12–11

$$p \, V \overline{(mw)} = z \, m \, RT$$

Replacing values gives

$$\left(111 \times 10^5\right) (0.002) \, (0.044) = \; z \, (0.560) \, (8.314)T$$

or

$$z \, T = 209.8$$

and multiplying and dividing by T_c gives

$$z \, T_r = \frac{zT}{T_c} = \frac{209.8}{304} = 0.69 \quad\quad\quad (12\text{–}14)$$

We solve this by trial and error.

Given p_r	Guess T_r	From Figure 12–9 z	Calculate zT_r and check with Eq. 12–14
1.5	1.1	0.49	0.54
1.5	1.2	0.68	0.82
1.5	1.15	0.60	0.69 ← *correct value*

So $T_r = 1.15$, and

$$T = T_r T_c = 1.15(304) = \underline{349.6 \text{ K}} = \underline{76.6°C} \quad \longleftarrow$$

Comment. If you check, you will see that our thermo tables (most others too) only deal with fluids below their critical point, thus in the region where liquid and gas can exist together. There is good reason for particular interest in this region because this is where by cleverly vaporizing and heating fluid on one path and then cooling and condensing it on another path, one can accomplish all sorts of magical things, such as extracting useful work, or doing refrigeration and air conditioning. As a hint of what can be done, look at paths ABC and ADC in problem 30.

Beyond the critical point we have to rely on methods based on gas laws to evaluate the properties of fluids. We do not treat these here (this is treated in any number of thermo texts). Finally, let us point out that thermo charts such as shown in Figure 12–10 span the whole range of conditions, relating the properties of

$$p, v, T, \text{ and } h$$

for vapor, liquid and the supercritical regime. However, unless one uses extra large 2 m × 2 m charts, they are usually not accurate enough for engineering use.

EXAMPLE 12–9.   An exploding liquid filled bottle

A 2-liter plastic pop bottle is filled with liquid water at 10 bar and at its boiling point. Surroundings are at 1 bar. The bottle explodes. How much work does it do?

Solution

The amount of water we are dealing with is

$$m = \frac{\text{volume of bottle}}{v_\ell, \text{ m}^3/\text{kg}} = \frac{0.002}{0.001\ 127} = 1.7746 \text{ kg}$$

Following the reasoning of Example 11–6, we conclude that it most reasonable to assume that an explosion is an adiabatic highly irreversible process, in which case we can write

$$W = -\Delta U = m(u_1 - u_2) \tag{12–15}$$

initial↘   ↙*final state*

and

$$W = m\, p_{\text{surr}}(v_2 - v_1) \tag{12–16}$$

Now at 10 bar the boiling point of the water is 179.91°C, but at 1 bar the boiling point is 99.63°C. So in dropping down to 1 bar the liquid has to cool from 179.91°C to 99.63°C or lower, otherwise it will partly or completely vaporize. Let us guess partial vaporization, with quality x. Then Eqs. 12–15 and 12–16 become

$$W = m\left[u_1 - (1-x)\, u_{2\ell} - x\, u_{2g}\right]$$

$$W = m\, p_{\text{surr}}\left[(1-x)\, v_{2\ell} + x\, v_{2g} - v_1\right]$$

Replacing values from the appendix gives for Eq. 12–15

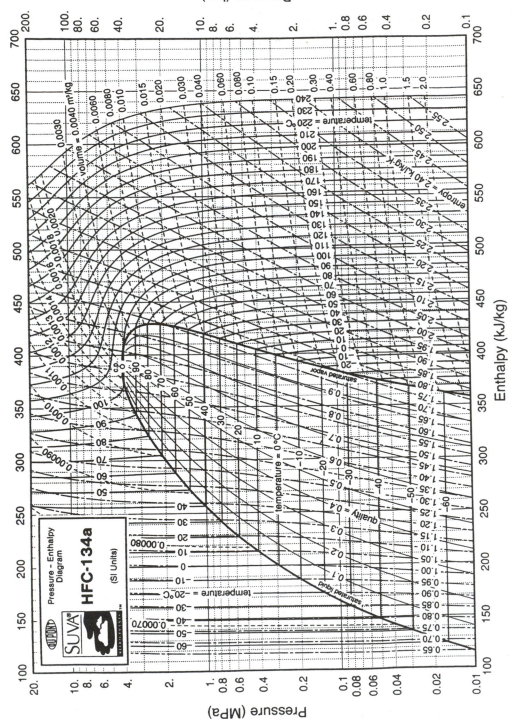

Figure 12–10. The p-h diagram for HFC-134a. From DuPont Technical Bulletin P-134a-SI, with permission.

$$W = 1.7746 \; \overbrace{[761\,680 - (1-x)417\,360 - x(2506\,100)]}^{u_2}$$

$$= 611\,030.3 - 3706\,678\,x \; \text{J} \tag{12–17}$$

and also for Eq. 12–16

$$W = 1.7746 \times 10^5 [\underset{\uparrow}{(1-x)(0.001\,043)} + \overbrace{x(1.694) - 0.001\,127}^{v_2}]$$

$$\underset{P_{surr}}{} \qquad$$

$$= -14.9066 + 300\,432x \; \text{J} \tag{12–18}$$

Solving Eqs. iii and iv simultaneously gives

$$\text{Vapor fraction:} \qquad x = 0.1525$$

$$\text{Liquid fraction:} \qquad 1 - x = 0.8475$$

and the work done by the explosion, from either Eq. 12–17 or Eq. 12–18 is

$$W = \underline{\underline{45\,798\ \text{J}}} = \underline{\underline{46\ \text{kJ}}} \quad \longleftarrow$$

PROBLEMS

1. Lemon-quench, a popular Balinese drink, contains 25% of sucrose by weight, $C_{12}H_{22}O_{11}$, the rest is mainly water. Find the mole fraction of sugar in the drink.

2. What is the density of a gaseous mixture of 70 vol% propane and 30 vol% methane at 3 atm and 200°C?

3. A process gas (10^6 m³/hr at 1.3 atm and 400°C) has the following molar composition

$$CH_4 - 15\%, \qquad C_6H_6 - 40\%, \qquad H_2 - 45\%$$

This gas is cooled and passed through an absorber where benzene is removed completely. Gas leaves the absorber at 30°C and 1 atm.

(a) Find the volumetric composition of gas leaving the absorber.

(b) Find the flow rate of gas in m³/hr leaving the absorber.

(c) Find the mean molecular mass of this leaving gas.

(d) Find the mass flow rate of this leaving gas.

4. Methanol is burned to completion with a stoichiometric amount of tired air,[3] or 20% oxygen and 80% nitrogen.

    (a) How many liters of air at 60°C and 103.82 kPa are needed to burn 1 mol of methanol?

    (b) What is the % composition of the resulting flue gas measured at 156°C and 720 mm Hg?

5. Three kilograms of anthracite, or hard coal (80% carbon, 20% hydrogen . . . by mass, of course), are burned to completion using a stoichiometric amount of tired air (20% O_2, 80% N_2 . . . by volume, of course).

    (a) How many moles of air are needed?

    (b) How many m^3 of combustion gases at 200°C and 700 mm Hg are produced?

    Suggestion: Take the 3 kg of coal as a basis for your calculations.

6. What fraction of liquid is represented by point d in Figure 12–6?

7. What phases are represented by points u, v, w, x, y, and z in Figure 12–5?

8. What phases are represented by points a, b, c, and d in Figure 12–5?

9. A 2.5 liter bomb contains nothing but 0.5 kg of water at 370°C. How much liquid does it contain?

10. At what temperature and pressure does boiling liquid water have the same density as atmospheric ice at –4°C?

11. Our refrigerator uses HFC-134a as its working fluid. In one part of the circuit, liquid refrigerant boils at −10°C. What is the molar latent heat of the fluid at this condition?

12. A rigid sealed 160-liter vessel contains 2 kg of water at room temperature (no air). This water is heated until the pressure in the vessel is 40 bar. What is the temperature of the water at this condition? Is it liquid, vapor or a mixture? If a mixture, what is its quality?

13. I asked that outfit at Albany to refill our tank of HFC-134a and their price was so low I now wonder whether they really filled it with liquid. Maybe it is only part liquid, or maybe only just gas at high pressure. Unfortunately, I didn't weigh the tank before and after filling, but the pressure gauge reads

---

[3]We will use the term tired air in some of our problems because it gives nice numbers.

10 atm and the tank has been sitting in our laboratory since Monday. What's your opinion—is it all liquid or did they cheat us?

In the following situations what does the vessel contain—gas, liquid, solid, or a mixture of phases? If it is a two-phase mixture, what is its quality? In all cases, what is the temperature of the container?

14. . . . A 20-liter vessel that contains 50 gm of water at 10 bar

15. . . . A 20-liter vessel that contains 17.8 kg of water at 50 bar

16. . . . A 20-liter vessel that contains 200 gm of water at 10 bar

---

17. One gram of HFC-134a is injected into a 10 cm^3 evacuated vessel, which is then heated to 30°C.

   (a) What is the mass fraction of vapor and of liquid in the vessel?

   (b) What is the volume fraction of vapor and of liquid in the vessel?

   (c) What is the pressure in the container?

18. Our laboratory plans to regularly use one 10-liter tankful of HFC-134a every week. So off I go to Albany to check the cost of fill-ups at the two possible suppliers.

    • At Chemical Mart the filled tank is returned to me at 10 bar and 90°C for $10.00.

    • At Super Discount Stores they also wanted to charge me $10.00 for the fill-up. I protested that the charge should be only $6.00 since they only gave me 6 bar at 0°C. We bargained and they eventually lowered the price to $8.00.

I'm not happy about this; however, to be scientific about it all, let's calculate the cost/kg charged by these two suppliers and then decide who to buy from.

19. From tables estimate the specific heat, c_p, of HFC-134a at 10 bar and in the temperature range of 100°C to 140°C.

20. Find the temperature and enthalpy of steam at 5 MPa and having a quality of 50%.

21. I have here a 1-liter tank of HFC-134a at 20°C and 0.1 MPa. What is the internal energy

   (a) per kg of the tank's contents?

   (b) of all the material in the tank?

22. A cylinder with a piston contains 2 kg of water at 20°C and 3 bar, but nothing else. While the pressure is kept constant at 3 bar, heat is added until the water is at 300°C. Find the work done by the water as it expands and pushes back the piston.

23. Steam in a piston-cylinder assembly at 900°C and 1.4 MPa is cooled at constant pressure to 300°C. How much heat must be removed?

24. Liquid HFC-134a at −10°C and 201 kPa is heated at constant pressure to 20°C.

(a) What is the heat input?

(b) How much work is done?

25. Steam in a cylinder expands adiabatically from 900°C and 2.0 MPa to 600°C and 0.4 MPa. How much work does it do?

We need to evaluate the density of the following fluids. Would you use our thermodynamic tables, ideal gas law, or the compressibility factor correction? What do you find?

26. . . . Propane at 63.9 bar and 425 K.

27. . . . Carbon monoxide at 39 bar and 1100°C.

28. . . . HFC-134a at 41.4 bar and 100°C.

29. . . . Water at 3 bar and 133.55°C.

---

30. In the drawing for Example 5 we showed a number of different ways of getting from state A to state C. For a 1-kilogram batch of HFC-134a

(a) calculate the work needed for path ABC.

(b) calculate the work needed for path ADC.

(c) If you were planning to design a circulating system for producing work, which path would you choose for the fluid, ABCDABCDA . . . or ADCBADCBA . . . ?

31. Repeat Example 9 with one change: The bottle is filled with saturated steam at 10 bar, not saturated liquid water. By the way, which do you think will give a bigger "bang," the exploding liquid or the exploding vapor?

32. Repeat Example 9 with one change: The liquid explodes at 99.63°C and 50 bar.

13
STEADY STATE FLOW SYSTEMS

Consider the flow systems shown in Figure 13–1.

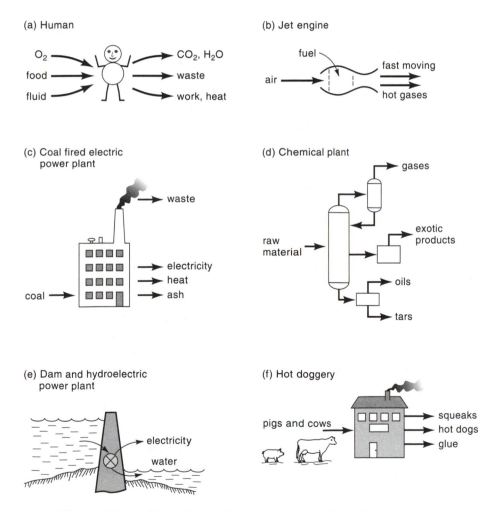

(a) Human

O_2 ⟶
food ⟶
fluid ⟶
⟶ CO_2, H_2O
⟶ waste
⟶ work, heat

(b) Jet engine

fuel
air ⟶
⟶ fast moving
⟶ hot gases

(c) Coal fired electric power plant

coal ⟶
⟶ waste
⟶ electricity
⟶ heat
⟶ ash

(d) Chemical plant

⟶ gases
raw material ⟶
⟶ exotic products
⟶ oils
⟶ tars

(e) Dam and hydroelectric power plant

⟶ electricity
water

(f) Hot doggery

pigs and cows ⟶
⟶ squeaks
⟶ hot dogs
⟶ glue

Figure 13–1. Examples of flow systems, which in the proper time scale can be considered to be steady state flow systems.

These examples may be analyzed as steady state or nonsteady state systems depending on the time scale considered. As an example, for the human, if we take a time scale in years, then the person gets wrinkles or gains weight, a child grows, a student learns thermo. At the other extreme, for very short times, say between 7:00 and 8:00 PM, before and after dinner a person's weight will go up by the food eaten. Actually, a person's mass will change with every breath taken (Figure 13–2).

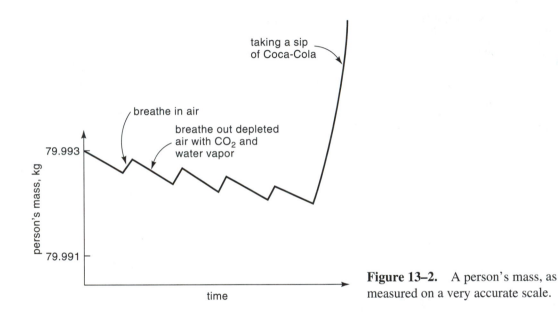

Figure 13–2. A person's mass, as measured on a very accurate scale.

There are all sorts of fluctuations in a person's mass; however, in the intermediate time scale, say days or weeks, we can reasonably assume an unchanging person, hence can treat and approximate that person as a steady state flow system.

In choosing the proper time scale we can reasonably approximate the other examples of Figure 13–1 as steady state flow systems. Let us next develop the necessary energy equations for such systems. For a system unchanging with time (Figure 13–3)

$$\text{all energy inputs} = \text{all energy outputs}$$

In a time interval Δt we write this as

$$E_1 + Q = E_2 + W \qquad [\text{J}]$$

$\underset{entering}{\underline{\quad}} \qquad \underset{leaving \; stream}{\underline{\quad}}$

or

$$(\Delta U + \Delta E_p + \Delta E_k)_{\text{streams}} = Q - W \qquad [J]$$

$$\underbrace{\qquad\qquad\qquad\qquad}_{\Delta E}$$

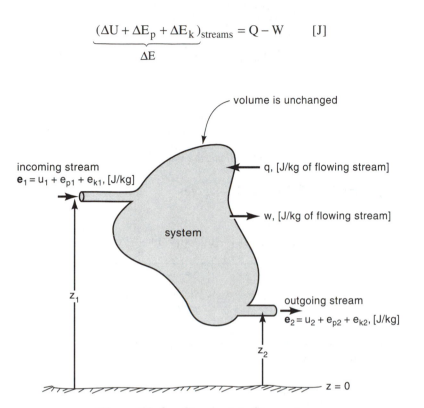

volume is unchanged

incoming stream
$e_1 = u_1 + e_{p1} + e_{k1}$, [J/kg]

q, [J/kg of flowing stream]

w, [J/kg of flowing stream]

system

z_1

outgoing stream
$e_2 = u_2 + e_{p2} + e_{k2}$, [J/kg]

z_2

z = 0

Figure 13–3.   Steady state flow system.

Based on unit mass of material entering (or leaving) we can then write

total work done
by system

$$(u_2 + e_{p2} + e_{k2}) - (u_1 + e_{p1} + e_{k1}) = q - w$$

or

$$\Delta u + \Delta e_p + \Delta e_k = q - w \qquad [J/kg] \qquad (13\text{--}1)$$

But for each unit mass leaving the system a volume v_2 leaves at p_2, pushes back the atmosphere and does work on the surroundings, or

$$w_{\text{pv,out}} = \int_0^{v_2} p \, dv = p_2 v_2$$

Similarly, the pv work received by the system by entering fluid is $w_{\text{pv,in}} = p_1 v_1$. This means that simply having material enter and leave the system involves work, pv work. Thus

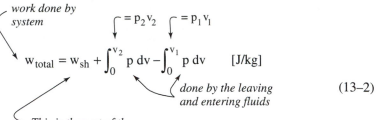

$$w_{total} = w_{sh} + \int_0^{v_2} p \, dv - \int_0^{v_1} p \, dv \qquad [\text{J/kg}] \qquad (13\text{–}2)$$

Replacing Eq. 13–2 in Eq. 13–1 gives

$$(u_2 + p_2 v_2) - (u_1 + p_1 v_1) + \frac{g}{g_c} \Delta z + \frac{\Delta v^2}{2 \, g_c} = q - w_{sh}$$

*for the flow streams,
not the system*

Again the enthalpy of the streams appears, so we have, in its various useful forms, for steady state flow systems,

per unit
of mass
flowing
$$\Delta h + \frac{g}{g_c} \Delta z + \frac{\Delta u^2}{2 \, g_c} = q - w_{sh} \qquad [\text{J/kg}]$$

per mol
flowing
$$\Delta h + \frac{(\overline{mw}) \, g \, \Delta z}{g_c} + \frac{(\overline{mw}) \, \Delta u^2}{2 \, g_c} = q - w_{sh} \qquad [\text{J/mol}] \qquad (13\text{–}3)$$

per unit
of time
$$\dot{m} \, \Delta h + \frac{\dot{m} \, g \, \Delta z}{g_c} + \frac{\dot{m} \, \Delta u^2}{2 \, g_c} = \dot{Q} - \dot{W}_{sh} \qquad [\text{J/s} = \text{W}]$$

Let us now consider some useful special case approximations of steady state flow systems.

EXAMPLE 13–1.   The steam or water turbine and the steam engine (Figure 13–4)

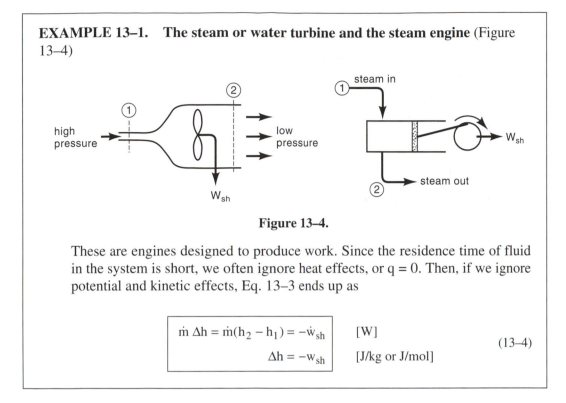

Figure 13–4.

These are engines designed to produce work. Since the residence time of fluid in the system is short, we often ignore heat effects, or q = 0. Then, if we ignore potential and kinetic effects, Eq. 13–3 ends up as

$$\dot{m}\,\Delta h = \dot{m}(h_2 - h_1) = -\dot{w}_{sh} \qquad [\text{W}]$$

$$\Delta h = -w_{sh} \qquad [\text{J/kg or J/mol}]$$

$$(13\text{–}4)$$

EXAMPLE 13–2.   The adiabatic flow nozzle (Figure 13–5)

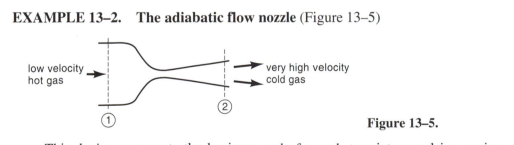

Figure 13–5.

This device represents the business end of a rocket or jet propulsion engine. Assuming $\Delta e_p = q = w_{sh} = 0$ and negligible incoming velocity, Eq. 13–3 becomes

$$\dot{m}\,\Delta h = -\frac{\dot{m}\mathbf{v}_2^2}{2\,g_c} \cdots \quad [\text{W}]$$

$$\Delta h = -\frac{\mathbf{v}_2^2}{2\,g_c} \cdots \quad \left[\frac{\text{J}}{\text{kg}}\right]$$

$$(13\text{–}5)$$

EXAMPLE 13–3.   The Joule-Thomson expansion (Figure 13–6)

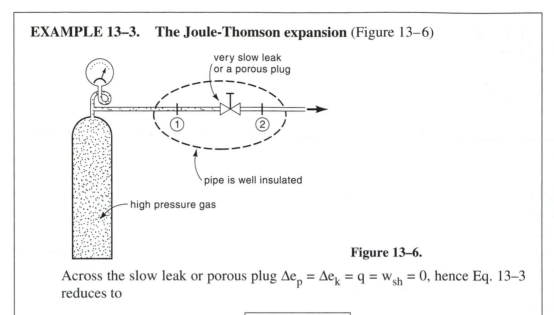

Figure 13–6.

Across the slow leak or porous plug $\Delta e_p = \Delta e_k = q = w_{sh} = 0$, hence Eq. 13–3 reduces to

$$\Delta h = h_2 - h_1 = 0$$

(13–6)

EXAMPLE 13–4.   The flow heater (Figure 13–7)

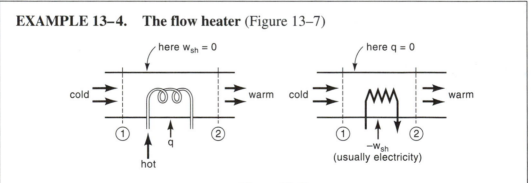

Figure 13–7.

This simple device represents the hot water heater in your home. In research it is used to find values for Δh. With $\Delta e_p = \Delta e_k = 0$ and with either $w_{sh} = 0$ or $q = 0$, Eq. 13–3 simplifies to

or

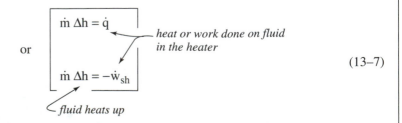

$$\dot m\, \Delta h = \dot q$$
— *heat or work done on fluid in the heater*

$$\dot m\, \Delta h = -\dot w_{sh}$$
fluid heats up

(13–7)

EXAMPLE 13–5.   Ideal piston-cylinder engine or ideal piston-cylinder pump
(Figure 13–8)

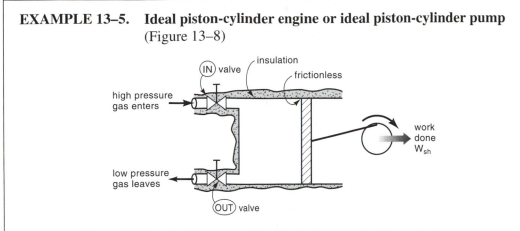

Figure 13–8.   Ideal or reversible (frictionless and adiabatic) piston-cylinder work-producing engine.

By ideal we assume that these devices are frictionless and well insulated. Thus they can work either as a pump to compress a gas, or as an engine to produce work from high pressure gas. Figure 13–8 shows the engine; for the pump, just reverse all three thick arrows.

   Let us analyze the engine. For this, determine the work obtained when we introduce 1 kg of high pressure gas in one cycle of the engine. Now this certainly is not a steady state operation, but if you look at a long-term operation of many cycles then you get the equivalent of a steady state operation of the engine (the system). Start with the cylinder empty, both valves closed, and follow the cycle on the pv diagram of Figure 13–9.

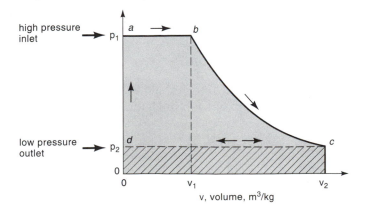

Figure 13–9.   Analysis of a piston-cylinder work-producing engine. Dotted area represents the work done by engine, hatched area represents work received from surroundings. Area a b c d is the net work done per cycle of the engine.

Step 1 Open the IN valve and introduce 1 kg of high pressure gas at p_1 and of volume v_1. Then close the IN valve. In Figure 13–9 we go from point a to point b. The work done by the gas and by the engine in this step is then

$$w_1 = \int_0^{v_1} p_1 dv = p_1 v_1 \qquad (13\text{–}8)$$

Step 2 Expand the gas to the outlet pressure p_2 (remember that both valves are closed), and this means going from point b to point c in Figure 13–9. The work done is then

$$w_2 = \int_{v_1}^{v_2} p \, dv \qquad (13\text{–}9)$$

Step 3 Open the OUT valve and push out all the gas in the cylinder (from point c to point d on Figure 13–9). The pressure is now p_2 so the work done by the engine (negative) is

$$w_3 = \int_{v_2}^0 p_2 \, dv = -p_2 v_2 \qquad (13\text{–}10)$$

this negative sign shows that this is not work done by the engine

Step 4 Shut the OUT valve. This completes the cycle.

Now the sum of these three work terms represents the net shaft work done by the fluid, which for one cycle is

$$w_{sh} = w_1 + w_2 + w_3$$

$$= p_1 v_1 + \int_{v_1}^{v_2} p \, dv - p_2 v_2 \qquad (13\text{–}11)$$

dotted area of Figure 13–9   *hatched area of Figure 13–9*

Subtracting the hatched area from the dotted area, we see that the net shaft work done per kg of fluid passing through the engine is

$$w_{sh} = -\int_{p_1}^{p_2} v \, dp \qquad \left[\frac{J}{kg}\right] \qquad (13\text{–}12)$$

Alternatively, we can come to the same conclusion using math. Thus integrating

$$d\,(pv) = pdv + vdp$$

gives

$$\int_1^2 d\,(pv) = p_2 v_2 - p_1 v_1 = \int pdv + \int vdp \qquad (13\text{–}13)$$

Replacing Eq. 13–13 into Eq. 13–11 gives Eq. 13–12.

So, with $\Delta e_p = \Delta e_k = q = 0$ and reversible operations Eq. 13–12 in Eq. 13–3 gives

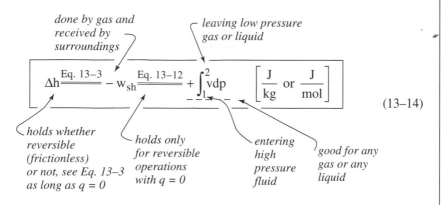

$$\Delta h \overset{\text{Eq. 13-3}}{=\!=\!=\!=} - w_{sh} \overset{\text{Eq. 13-12}}{=\!=\!=\!=} + \int_1^2 vdp \qquad \left[\frac{J}{kg} \text{ or } \frac{J}{mol}\right] \qquad (13\text{–}14)$$

done by gas and received by surroundings — *leaving low pressure gas or liquid* — *holds whether reversible (frictionless) or not, see Eq. 13-3 as long as q = 0* — *holds only for reversible operations with q = 0* — *entering high pressure fluid* — *good for any gas or any liquid*

We derived this expression for piston-cylinder operations of pumps or work-producing engines; however, it holds also for continuous operations of turbines or compressors and for any fluid, even peanut butter. This is an important equation, and is the starting point for the analysis of all internal combustion engines. Think of that when you drive home in your 50 mile/gal Honda.

EXAMPLE 13–6.   Ideal turbine or compressor (with either gas or liquid)
(Figure 13–10)

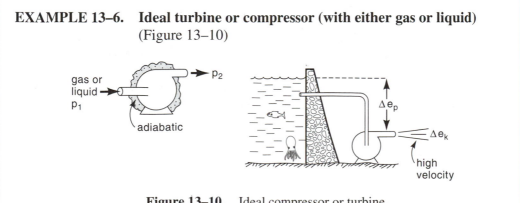

Figure 13–10.   Ideal compressor or turbine.

Start with Eq. 13–1 and for an ideal frictionless device $(q = 0)$

$$\Delta u + \Delta e_p + \Delta e_k = q - w = -\int pdv$$

Combining with Eq. 13–13 gives

$$\Delta u + \Delta e_p + \Delta e_k = -p_2 v_2 + p_1 v_1 + \int v \, dp$$

or

$$\boxed{\Delta h + \Delta e_p + \Delta e_k = +\int v \, dp} \qquad (13\text{--}15)$$

Note that Eq. 13–15 is essentially equivalent to Eq. 13–14.

For the special case of an incompressible liquid, v = constant, Eq. 13–15 reduces to

$$\boxed{\Delta h + \Delta e_p + \Delta e_k = v\Delta p = \frac{\Delta p}{\rho}} \qquad (13\text{--}16)$$

EXAMPLE 13–7.   Ideal isothermal turbine or compressor processing an ideal gas (Figure 13–11)

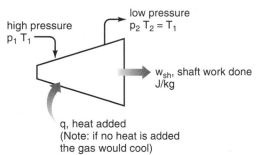

Figure 13–11.   Ideal isothermal work-producing machine

Assume that $\Delta e_p = \Delta e_k = 0$. Then noting that $T_1 = T_2$, $\Rightarrow \Delta h = 0$ and $p_1 v_1 = p_2 v_2$ Eq. 13–3 with Eq. 13–12 gives, per mole of flowing gas,

or

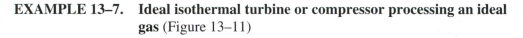

$$\boxed{\begin{aligned} w_{sh} &= q = -\int v \, dp = RT \, \ln \frac{p_1}{p_2} \quad \left[\frac{J}{mol}\right] \\[2mm] \dot{W}_{sh} &= \dot{Q} = \dot{n} \, RT \, \ln \frac{p_1}{p_2} \qquad [W] \end{aligned}} \qquad (13\text{--}17)$$

Here for isothermal operations we get the same equation for batch and flow systems (see Eq. 11–13). This is not so for adiabatic operations.

**EXAMPLE 13–8.   Ideal frictionless adiabatic turbine or compressor
processing an ideal gas** (Figure 13–12)

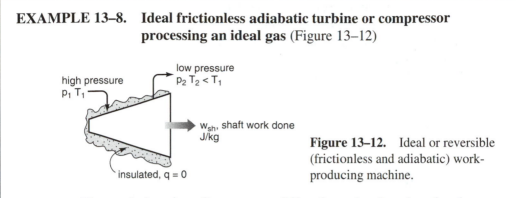

Figure 13–12.   Ideal or reversible
(frictionless and adiabatic) work-
producing machine.

The work done by a flow system differs from that done by a batch system
by the difference between Δu from Δh, or by Δpv. So per mole we can write

$$w_{flow} = w_{batch} - \Delta pv \tag{13-18}$$

For adiabatic reversible operations Eq. 11–17 with Eq. 13–18 gives

$$w_{flow} = -\frac{(p_2 v_2 - p_1 v_1)}{k - 1} - (p_2 v_2 - p_1 v_1)$$

$$= -\frac{k}{k - 1}(p_2 v_2 - p_1 v_1) = k w_{batch} \tag{13-19}$$

So with Eq. 11–17 for batch systems, in rate form, we get the power production as

$$\dot{W}_{sh} = \frac{k p_1 \dot{V}_1}{k - 1}\left[1 - \left(\frac{p_2}{p_1}\right)^{(k-1)/k}\right] = \dot{n} c_p T_1 \left[1 - \left(\frac{p_2}{p_1}\right)^{(k-1)/k}\right]$$

$$= -\dot{n} c_p (T_2 - T_1) = -\frac{k}{k - 1}(p_2 \dot{V}_2 - p_1 \dot{V}_1) \qquad [W] \tag{13-20}$$

EXAMPLE 13–9.   Real turbines and compressors (Figure 13–13)

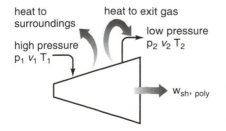

Figure 13–13.   Real turbine (or
compressor) has friction and heat
transfer effects.

Real turbines or compressors are not perfectly insulated and they also generate heat by friction. This may result in heat loss to the surroundings, and to heating of the exiting gases. The equations for this case are clumsy, so we won't consider them. However, for a *polytropic change* reasoning similar to that leading to Eqs. 13–18 and 13–19 gives

$$w_{sh,flow} = \gamma w_{batch} \tag{13-21}$$

Finally, Eq. 11–17 with Eq. 13–21 gives

$$\dot{W}_{sh} = \frac{\gamma p_1 \dot{V}_1}{\gamma - 1}\left[1 - \left(\frac{p_2}{p_1}\right)^{(\gamma-1)/\gamma}\right] = \frac{\gamma \dot{n}RT_1}{\gamma - 1}\left[1 - \left(\frac{p_2}{p_1}\right)^{(\gamma-1)/\gamma}\right]$$

$$= -\frac{\gamma \dot{n}R}{\gamma - 1}(T_2 - T_1) = -\frac{\gamma(p_2\dot{V}_2 - p_1\dot{V}_1)}{\gamma - 1} \qquad [W] \tag{13-22}$$

EXAMPLE 13–10.   Pumping up a tank with an ideal gas (Figure 13–14)

A 10 m³ tank is open to the surroundings at 20°C and 1 bar. A compressor is then connected to the tank's inlet, is switched on, and pumps air into the tank. The compressor operates isothermally.

    (a) Find the minimum work required to pressurize the tank to 10 bar.

    (b) Find the heat interchange at the compressor.

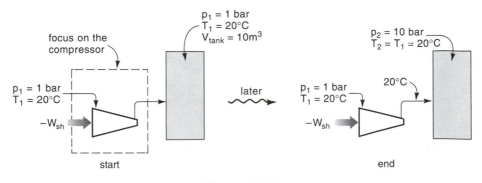

Figure 13–14.

Solution

If we look at the overall system, compressor and tank, we see that it is an unsteady state operation. However, if we focus on the compressor we can avoid treating this as an unsteady state problem, to be considered in the next chapter. So focus on the compressor.

From Example 13–7 we see that the work needed to compress dn moles of ideal gas from p_1 to p, all at T_1 is

$$dW_{sh} = \left(RT_1 \, \ell n \frac{p_1}{p} \right) dn \quad [J]$$

But the pressure in the tank changes from p_1 to p_2 as n increases. So relate p to n by the ideal gas law

$$n = \frac{V_{tank} \, P}{RT_1} \quad \text{or} \quad dn = \frac{V_{tank}}{RT_1} \cdot dp$$

Combining the above two equations gives

$$dW_{sh} = \frac{\cancel{RT_1} V_{tank}}{\cancel{RT_1}} \, \ell n \frac{p_1}{p} \, dp = V_{tank} \, \ell n \frac{p_1}{p} \, dp$$

or

$$\boxed{W_{sh} = V_{tank} \int_{p_1}^{p_2} \ell n \frac{p_1}{p} \, dp \cdots} \quad [J] \tag{i}$$

Introducing values from this problem

$$W = 10 \text{ m}^3 \int_{10^5}^{10^6} \ell n \frac{10^5}{p} \, dp$$

Solving graphically Figure 13–15 gives

$$W = (10 m^3)(-1.35 \times 10^6 \text{ Pa}) = -13.5 \times 10^6 J$$

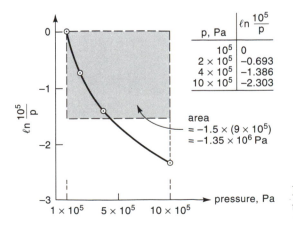

p, Pa	$\ell n \dfrac{10^5}{p}$
10^5	0
2×10^5	−0.693
4×10^5	−1.386
10×10^5	−2.303

area
$= -1.5 \times (9 \times 10^5)$
$= -1.35 \times 10^6$ Pa

Figure 13–15.   Graphical solution to Example 13–10.

Thus, the work required is

*Solving analytically gives 14.0
hence graphical integration is 3.7% low*

$$-W_{sh} = 13.5 \text{ MJ} \longleftarrow$$

(b) The heat to be removed during the compression is gotten from Eq. 13–3

$$m[\Delta h + \Delta e_p + \Delta e_k] = Q - W_{sh}$$

Thus, the heat to be removed is

$$-Q = 13.5 \text{ MJ} \longleftarrow$$

EXAMPLE 13–11.   The flow reactor

1 mol/s of gaseous A and 1 mol/s of gaseous B, both at 25°C, are pumped continuously into an adiabatic mixer-reactor. They react to completion according to the stoichiometry

$$A + B \rightarrow R$$

The product stream, also gaseous, leaves the reactor at 225°C. Find the ΔH_r for the above reaction at 525°C.

Data $c_{pA} = 30, \quad c_{pB} = 40, \quad c_{pR} = 50$ J/mol·K

Solution

Let us sketch the system (Figure 13–16)

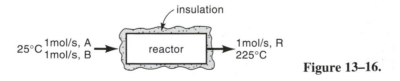

$$25°C \begin{array}{l} 1\text{mol/s, A} \\ 1\text{mol/s, B} \end{array} \rightarrow \boxed{\text{reactor}} \rightarrow \begin{array}{l} 1\text{mol/s, R} \\ 225°C \end{array}$$

insulation

Figure 13–16.

Then the first law expression, Eq. 13–3, from input to output stream becomes

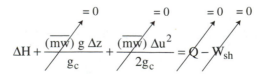

$$\Delta H + \cancel{\frac{(m\cancel{w}) \, g \, \Delta z}{g_c}}^{=0} + \cancel{\frac{(m\cancel{w}) \, \Delta u^2}{2g_c}}^{=0} = \cancel{Q}^{=0} - \cancel{W_{sh}}^{=0}$$

or

$$\Delta H = 0 \quad \text{or} \quad \sum \Delta H = 0$$

For the three-step path $0 \rightarrow 1 \rightarrow 2 \rightarrow 3$, Figure 13–17 shows that we have to account for ΔH for heating, reaction, followed by cooling.

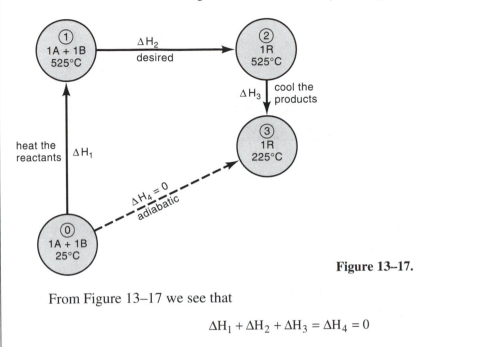

Figure 13–17.

From Figure 13–17 we see that

$$\Delta H_1 + \Delta H_2 + \Delta H_3 = \Delta H_4 = 0$$

or

$$\Delta H_2 = -\Delta H_1 - \Delta H_3$$

$$= -[1 \cdot c_{pA}(T_1 - T_0) + 1 \cdot c_{pB}(T_1 - T_0)] - 1 \cdot c_{pR}(T_3 - T_2)$$

$$= -[30(525 - 25) + 40(525 - 25)] - 50(225 - 525)$$

$$= -[15\ 000 + 20\ 000] + 15\ 000$$

Finally

$$\Delta H_2 = \Delta H_{r,525°C} = \underline{\underline{-20\ 000 \text{ J/mol A}}} \quad \longleftarrow$$

Note. Here the analysis involving drawing a flow map is analogous to that for a batch constant pressure system. So look up Chapter 9 for more examples.

This chapter developed the general equations for steady state flow equations. We then showed how they could be simplified to represent a number of special cases. Now it is time to try to solve some problems.

PROBLEMS

1. A stream of liquid n-hexane (C_6H_{14}, $c_p = 201.5$ J/mol·K) at about 1 atm is to be heated from 25°C to 68°C before entering a reactor. The flow rate to the reactor is to be 1 ton/hr. How many 1000 W heaters are needed to do the job?

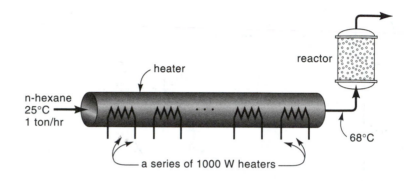

2. A stream of helium (10 ft³/min, 49°C, 2.8 atm) enters and flows through a heater, leaving at 97°C. What is the energy input rate (watts or kilowatts) needed for this operation?

3. Methane (CH_4, c_v = 2600 J/kg·K) leaves a storage tank at 125 kPa and 300 K at 40 lit/s. It flows through a heat exchanger and then on to a reactor. What should be the rate of heat addition in the heat exchanger if methane is to be heated to 400 K?

4. We plan to build a shower room and hot tub for the employees of our chemical plant. Cold water we have, but hot water is the problem. How about mixing waste superheated steam, at 150°C and 1 bar with cold water at 5°C to make nice hot water at 50°C? How much cold water will we need to mix with each kg of steam to get the desired hot water?

5. 1000 kg/hr of superheated steam at 300°C and 1 bar is needed as a feed to a heat exchanger. To produce it saturated steam at 1 bar is mixed with a source of very hot superheated steam at 400°C and 1 bar. The mixing unit is adiabatic. Calculate the needed flow rate of the two feed streams.

6. A flow of 1000 m³/min of carbon dioxide (c_p = 55 J/mol·K) at 1500°C and 1 atm is cooled in a long tube heat exchanger to 700°C by cold tap water at 8°C heated to 68°C.

   (a) How much cooling is needed, in watts?

   (b) Determine the water flow rate needed.

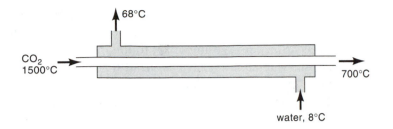

68°C

CO₂
1500°C

700°C

water, 8°C

7. A stream of 1 m³/s H_2O at 600°C and 6 bar is to be brought to liquid water at 85°C and 1 bar. What is the energy removal rate from this flowing stream? Please use steam tables to solve this problem.

8. 1000 kg/s of compressed water enters a hydraulic turbine at 10 bar. This water does work and then leaves at 1 bar. The inlet and outlet ports of the turbine are at the same elevation, the water velocities are small and the turbine casing is well insulated. In steady state operations calculate the power output of the turbine. Water enters and leaves at 8°C.

9. Here in the chemical engineering department we will do some experiments in a bed of fine solids suspended by upflowing gas. We call such a system a "fluidized bed." We plan to draw in air from the outside, compress it to 1.5 atm and 60°C, then heat it to 400°C. This air then is introduced into the fluidized bed at a superficial velocity of 4 m/s. The cross-sectional area of bed is 1.0 m².

(a) How many moles of air do we need per second?

(b) How much heat must each mole of air absorb in the heater, in J?

(c) If we use electrical heaters for the heating job, and if electricity costs 7¢/kW·hr, what would be cost of electricity for a 4 hr experimental run?

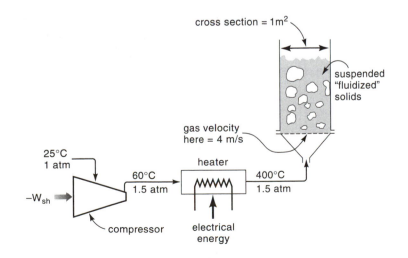

cross section = 1m²

suspended "fluidized" solids

gas velocity here = 4 m/s

25°C
1 atm

heater

60°C
1.5 atm

400°C
1.5 atm

$-W_{sh}$

compressor

electrical energy

10. *Artificial hearts.* The human heart is a wondrous pump, but only a pump. It has no feelings, no emotions, and its big drawback is that it only lasts one lifetime. Since it is so important to life, how about replacing it with a compact, super reliable mechanical heart that will last two lifetimes? Wouldn't that be great? The sketch below gives some pertinent details of the average relaxed human heart. From this information calculate the power requirement of an ideal replacement heart to do the job of the real thing.

Comment.   Of course, the final unit should be somewhat more powerful, maybe by a factor of five, to account for pumping inefficiencies and to take care of stressful situations, such as running away from hungry lions. Also assume that blood has the properties of water.

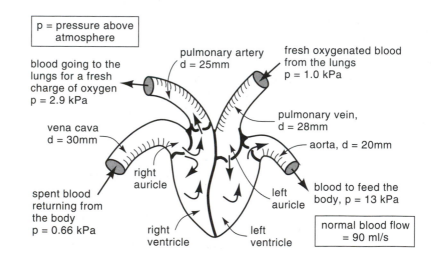

11. A storage tank contains a high pressure ideal gas at 100 bar and 300 K. The valve at the top of the tank has a slight leak. What is the temperature of the gas leaving the *valve*, T_3, when the pressure in the tank has dropped to 10 bar? The whole setup of tank and valve are well insulated. Assume ideal gas behavior.

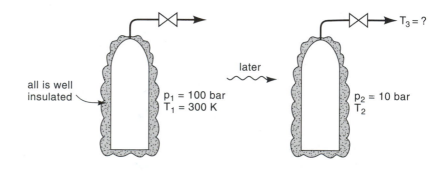

12. HFC-134a is flowing at 80°C and 20 bar from here to there in a pipeline. A small leak develops at a joint in the pipeline and this refrigerant leaks into the room. What is the temperature of this leaking gas if pipeline and valve are both well insulated?

13. The laboratories of the East Tibetan Institute of Technology are much like any other—machines, instruments, piping, etc.—but I have a question. Because of the high altitude I wonder if it is pure steam (100% quality) flowing through those steam pipes, which are at 1 MPa. So I crack a valve leading from the steam line and measure the temperature of the exiting stream and find it to be exactly 100°C.

What is the quality of the H_2O in the line; or if it is superheated, then by how many °C? By the way, the air pressure at the institute is 50 kPa, not 100 kPa.

14. Our laboratory (at 21.3°C and 1 bar) has need for a continuous stream of high pressure air (100 mol/s at 5 bar) and I've been asked to order a compressor for this job. New ones are too expensive, so I'm off to the second-hand reconditioned outlet that sells 50% efficient units compared to an adiabatic reversible unit.

(a) What size unit should I look for?

(b) What will be the temperature of the outlet air if the unit is well insulated?

15. The only place on earth where helium is found is in Texas natural gas. There it is separated, purified, compressed, and stored in large reservoirs. If 100 mol/s of helium are compressed continuously from 100 kPa to 1 MPa using a 94.3% efficient adiabatic compressor, how much power is needed to drive the compressor? Also determine the temperature of the leaving helium stream if it enters the compressor at 300 K.

16. One kg/s of fluid flows at a steady rate into and out of a system. The enthalpy, velocity, and height of the entering streams are 400 kJ/kg, 100 m/s, and 300 m. For the exit streams these quantities are 396 kJ/kg, 1 m/s, −10 m. Also 2037.5 W of heat are removed from the system. At what rate does the system do work, or at what rate is work done on the system?

17. In Example 10 the pump was assumed to operate isothermally. But this usually is not reasonable. So repeat Example 10 but assume that the compressor operates adiabatically reversibly and that the tank is well insulated.

18. Given the reaction:

$$A + B \rightarrow C, \qquad \Delta H_{r,25°C} = -32\,478 \text{ J}$$

All reaction components, A, B, C, are gases between 25°C and 1000°C, and their mean specific heats are

$$c_{pA} = 24 \text{ J/mol·K} \qquad c_{pB} = 30 \text{ J/mol·K} \qquad c_{pC} = 54 \text{ J/mol·K}$$

(a) Find the heat of reaction at 1000°C.

(b) Determine ΔU_r, the internal energy change, for this reaction at 25°C.

---

1 mol/s of gaseous A and 2 mol/s of gaseous B, both at 25°C, are pumped continuously into an adiabatic mixer-reactor. They react to completion according to the stoichiometry

$$A + B \rightarrow R$$

The product stream, also gaseous, leaves at 225°C. Find the heat of reaction for the reaction $A + B \rightarrow R$

19. ... at 25°C.

20. ... at 225°C.

21. ... at 0°C.

Data $c_{pA} = 30 \text{ J/mol·K}$
$c_{pB} = 35 \text{ J/mol·K}$
$c_{pR} = 40 \text{ J/mol·K}$

---

22. An equimolar mixture of A and B enter a flow reactor and react to completion according to the reaction $A + 2B \rightarrow 2R$. Per mole of entering A at 25°C, how much heat must we add or remove from the reactor if the product stream is to leave at 325°C?

Data $c_{pA}(g) = 40 \text{ J/mol·K} \qquad c_{pB}(g) = 45 \qquad c_{pR}(g) = 30$

$$c_{pB}(\ell) = 35 \qquad c_{pR}(\ell) = 50$$

$$A(g) + 2B(\ell) = 2R(\ell) \ldots \Delta H_{r,298K} = 120\,000 \text{ J}$$

$$\Delta H_{\ell g}(B) = 60\,000 \text{ J/mol at } T_b = 225°C$$

$$\Delta H_{\ell g}(R) = 30\,000 \text{ J/mol at } T_b = 125°C$$

23. Fine sulfur dust and oxygen, both at 25°C enter a reactor where the sulfur is burned. All the sulfur forms SO_2 and the exit gases are at 1000°C. How many moles of oxygen are we adding with each gram-atom (32 gm) of sulfur?

Data Between 25°C and 1000°C

$$\begin{cases} c_{p(SO_2)} = 50 \text{ J/mol·K} \\ c_{p(O_2)} = 32 \text{ J/mol·K} \\ c_{p(S_{solid})} = 35 \text{ J/mol·K} \end{cases}$$

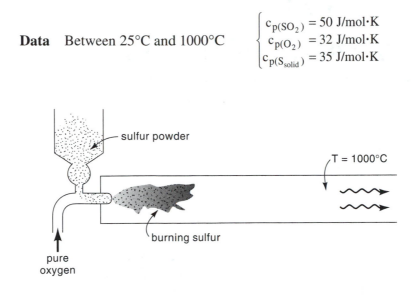

24. 1 mol/s of gaseous A, 1 mol/s of gaseous B, and 1 mol/s of inert carrier gas I are fed at 25°C to an adiabatic flow reactor where A and B react to completion to give R which is liquid at 25°C. However, reaction is exothermic and the product stream leaves the reactor at 325°C. Find $\Delta H_{r,25°C}$ for the reaction

$$A(g) + B(g) \rightarrow R(\ell)$$

Data For A: $c_p(g) = 30 \text{ J/mol·K}$ R melts at $-25°C$

For B: $c_p(g) = 40 \text{ J/mol·K}$ R boils at $125°C$

For R: $c_p(g) = 60 \text{ J/mol·K}$ $\Delta H_{\ell g} = 10\,000 \text{ J/mol}$

$c_p(\ell) = 50 \text{ J/mol·K}$

For I: $c_p(g) = 30 \text{ J/mol·K}$

25. Repeat the previous problem with one change. The feed stream consists of 1 mol A/s, 2 mol B/s, 1 mol inert/s.

CHAPTER 14
UNSTEADY STATE FLOW SYSTEMS

This chapter treats the most general case of the first law, the unsteady state flow system (that which changes with time). These are systems that either grow, or shrink, or change in composition. They do not stay unchanged with the passage of time. Let us consider some examples in Figures 14–1, 14–2, and 14–3.

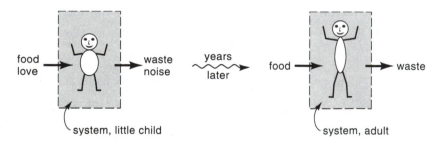

Figure 14–1. Human growth.

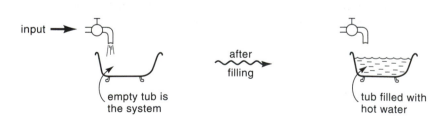

Figure 14–2. Filling a bathtub.

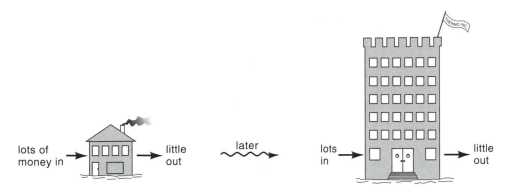

Figure 14–3. For a money analogy consider a growing profitable business.

In terms of energy we sketch these situations in Figure 14–4.

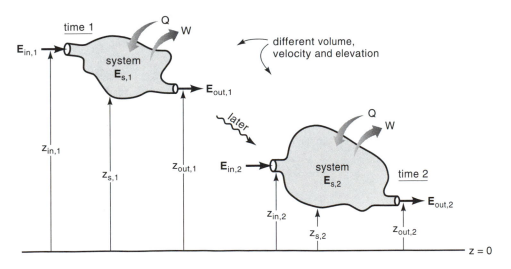

Figure 14–4.

To develop the corresponding equations, consider the changes in the energy of the system in a time interval between t_1 and t_2, or Δt

$$\Delta E_{system} = (\text{all energy inputs}) - (\text{all energy outputs})$$

$$= -\Delta E_{streams} + Q - W$$

or

$$\Delta E_{systems} + \Delta E_{streams} = Q - W \qquad (14\text{–}1)$$

In the time interval when the mass of the system goes from m_1 to m_2, when m_{in} has entered and m_{out} has left, and for flow rates that do not change with time, Eq. 14–1 can be written as

$$\underbrace{m_2(u + e_p + e_k)_2 - m_1(u + e_p + e_k)_1}_{system} + \underbrace{m_{out}(u + e_p + e_k)_{out} - m_{in}(u + e_p + e_k)_{in}}_{streams} = Q - W$$

$$(14\text{–}2)$$

where

$$
\begin{array}{c}
\overset{\displaystyle \text{\textit{if system expands it must}}}{\underset{\displaystyle \text{\textit{push back the atmosphere}}}{}} \\
W = W_{sh} + W_{pV,system} + W_{pV,streams} \qquad (14\text{–}3) \\
\underset{\textit{shaft work}}{} \qquad \underset{= (mpv)_{out} - (mpv)_{in}}{}
\end{array}
$$

Combining u and pv for the flow streams to give enthalpies (as was done in the previous chapter, see Eq. 13–3) we end up, for time interval Δt,

$$
\boxed{
\begin{aligned}
m_2(u + e_p + e_k)_2 &- m_1(u + e_p + e_k)_1 + m_{out}(h + e_p + e_k)_{out} - m_{in}(h + e_p + e_k)_{in} \\
&= Q - W'_{sh} \\
&= Q - (W_{sh} + W_{pV,system}) \\
&= Q - \left(W_{sh} + \int_{V_1}^{V_2} p_{system} dV \right) \qquad [J]
\end{aligned}
}
$$

change in system's volume

$$(14\text{–}4)$$

This expression applies to situations with unchanging flow rates.

For multiple entering and/or leaving flow streams, in the time period where the mass of the system goes from m_1 to m_2 we write

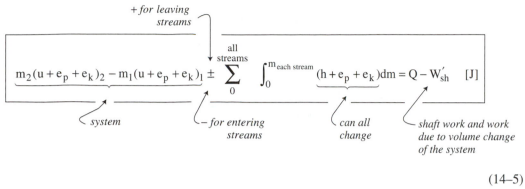

$$\underbrace{m_2(u+e_p+e_k)_2 - m_1(u+e_p+e_k)_1}_{\text{system}} \underset{\underset{-\text{for entering streams}}{\pm}}{\pm} \underbrace{\sum_0^{\text{all streams}} \int_0^{m_{\text{each stream}}} \overbrace{(h+e_p+e_k)}^{\text{can all change}} dm}_{} = Q - \overbrace{W_{sh}'}^{\text{shaft work and work due to volume change of the system}} \quad [J]$$

+ for leaving streams

(14–5)

The following examples illustrate the use of these equations.

EXAMPLE 14–1.   Filling a glass with water (Figure 14–5)

Hot water (80°C) from a kettle is poured into a completely insulated styrofoam cup. Apply the general equation to the cup to find the temperature of the water in the cup.

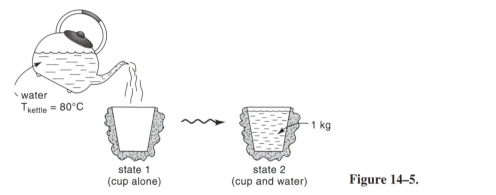

water
$T_{\text{kettle}} = 80°C$

— 1 kg

state 1
(cup alone)

state 2
(cup and water)

Figure 14–5.

Solution

As a basis let us consider one kilogram of hot water poured into the cup. Then Eq. 14–4 becomes

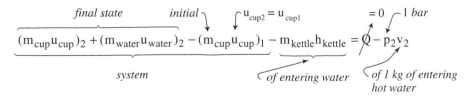

$$\underbrace{(m_{cup}u_{cup})_2 + (m_{water}u_{water})_2 - (m_{cup}u_{cup})_1}_{\text{system}} - \underbrace{m_{kettle}h_{kettle}}_{\text{of entering water}} = \cancel{Q} - \underbrace{p_2 v_2}_{\begin{array}{c}\text{of 1 kg of entering}\\\text{hot water}\end{array}}$$

final state   *initial* $u_{cup2} = u_{cup1}$ $= 0$ 1 bar

where $p_2 v_2$ represents the work done by the system as it pushes back the atmosphere. Simplifying gives

$$u_{water,2} - h_{kettle} = -p_2 v_{water,2}$$

or

$$h_{water,2} = h_{kettle}$$

Therefore

$$T_{water,2} = T_{kettle} = \underline{80°C} \longleftarrow$$

EXAMPLE 14–2.   Filling an evacuated tank with an ideal gas (Figure 14–6)

A valve on an evacuated insulated tank is opened. Air (an ideal gas) rushes in and the pressure equalizes. The valve is then quickly closed. What is the temperature of the gas in the tank if room temperature is 27°C and pressure is 1 bar?

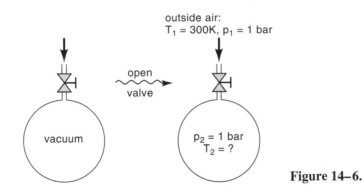

outside air:
$T_1 = 300K$, $p_1 = 1$ bar

open valve

vacuum

$p_2 = 1$ bar
$T_2 = ?$

Figure 14–6.

Assume $c_p = 29.1$ J/mol·K at all temperatures.

Solution

Let the gas in the tank be the system, and since the whole operation is done very quickly we can reasonably assume that it occurs adiabatically. Then per mole of air entering, Eq. 14–4 becomes

$$m_2(u + \cancel{e_p} + \cancel{e_k})_2 - \cancel{m_1(u + e_p + e_k)_1} + \cancel{0} - m_{in}(h + \cancel{e_p} + \cancel{e_k})_{in} = \cancel{Q} - \cancel{W'_{sh}} \qquad (14–6)$$

$= 0$, $= 0$ $= 0$, *vacuum* $m_{out} = 0$

$= m_2$

$$\therefore u_2 = h_{in}$$

But for an ideal with gas with c_p independent of temperature Eq. 11–7 combined with Eq. 14–6 gives

$$c_v T_2 = c_p T_{in}$$

or

$$T_2 = kT_{in} = \left(\frac{29.1}{29.1 - 8.314}\right)(300) = 420 \text{ K} = \underline{147°C} \quad \longleftarrow$$

This says that the cool gas becomes very hot as it enters the tank.

EXAMPLE 14–3. Topping a tank (Figure 14–7)

An extension of the previous example has some fluid originally in the tank, as sketched below.

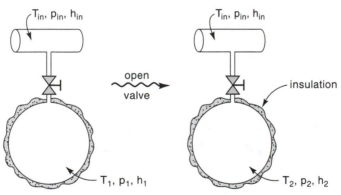

Figure 14–7.

Inserting the quantities into Eq. 14–4 and then simplifying gives

- for ideal gases: $T_2 = kT_{in} \Big/ \left[1 - \dfrac{p_1}{p_{in}} + k\dfrac{p_1}{p_{in}}\dfrac{T_{in}}{T_1}\right]$ $\longleftarrow$ (14–7)

- in general: $m_2 u_2 - m_1 u_1 - (m_2 - m_1)h_{in} = 0$ $\longleftarrow$ (14–8)

In Eq. 14–8 m_2 and u_2 are unknown. Guess T_2 and solve by trial and error using the tables.

EXAMPLE 14–4.   An early atmosphere experiment (Figure 14–8)

Theory has it that Earth's atmosphere in the carboniferous period (300 million years ago) was at a much higher pressure than it is today, and that it consisted primarily of CO_2. Plants loved it, gorged themselves on the CO_2, grew profusely worldwide, died, and thereby produced the vast coal deposits that we see today.

We are running an experiment today to check this theory by growing plants in a perfectly insulated high pressure vessel containing 1 m³ of 96% CO_2 and 4% O_2 at 5 bar and 27°C. To clean the unit we want to lower the pressure to 1 bar by opening the valve. But this would cool the gas (adiabatic reversible expansion, see Example 11–1). To counter this we switch on a 150 W light when we let the gas out. A controller adjusts the flow rate of gas so as to keep the temperature unchanged.

(a) To keep the vessel at 27°C what should be the flow rate of gas from the vessel?

(b) How long would it take us to lower the pressure from 5 bar to 1 bar?

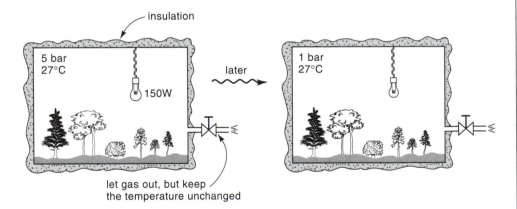

Figure 14–8.

Solution

Consider the vessel to be the system, sub s, and ignore e_k and e_p. Then Eq. 14–5, in molar terms, reduces to

$$V = const$$

$$n_2 u_2 - n_1 u_1 + h_{out} \underbrace{\int_0^{n_1 - n_2}}_{equal} dn = \cancel{Q} - W_{sh} - \cancel{W}_{pV}$$

$$\underbrace{(n_2 - n_1) u_s}_{negative} + \underbrace{h_{out} (n_1 - n_2)}_{positive} = -W_{sh}$$

$$(h_{out} - u_s) \, \Delta n = -W_{sh}$$

$$\swarrow leaving$$

In differential form this becomes

$$(h_{out} - u_s) \frac{d \, n_{out}}{dt} = -\frac{d \, W_{sh}}{dt}$$

$$(c_p - c_v) \, T \, \dot{n}_{out} = -\dot{W}_{sh}$$

$$\therefore \dot{n}_{out} = \frac{-\dot{W}_{sh}}{RT}$$

$$= \frac{150 \, W}{(8.314)(300)} = \underline{\underline{0.060 \, mol \, / \, s}} \quad \longleftarrow \quad (a)$$

To calculate the time needed to lower the pressure first determine the original number of mols of gas in the vessel

$$n_1 = \frac{p_1 \, V}{R \, T} = \frac{(5 \times 10^5) \, (1)}{(8.314) \, (300)} = 200 \, mol$$

and when the pressure has dropped to 1 bar

$$n_2 = 40 \, mol$$

So the time needed to lower the pressure is

$$t = \frac{\Delta n}{\dot{n}_{out}} = \frac{200 - 40}{0.060} = \underline{\underline{2667 \, s = 44 \, min \, 27 \, s}} \quad \longleftarrow \quad (b)$$

EXAMPLE 14–5.   To heat a cold exhibition hall

A large unused exhibition hall (50 m × 40 m × 10 m) is to be prepared for a show and has to be heated from 0°C to 27°C. How many 1.5 kW portable heaters operating for 24 hrs would be needed for this job?

Note   The pressure stays at 1 bar during the heating process, air leaks out of the hall, so the amount of air that needs to be heated changes with time. Only account for the heating of air, not walls, fixtures, and furniture. This is a terrible assumption, but . . .

Solution

We present three solutions: a rigorous, an approximate, and a clever simple solution.

Method 1.   Rigorous solution

For the air in the exhibition hall Eq. 14–5 in molar units becomes

$$n_2 u_2 - n_1 u_1 + \int^{n_1 - n_2} h_{out} dn = Q$$

or

$$n_2 c_v T_2 - n_1 c_v T_1 + \int_0^{n_1 - n_2} c_p \underset{\underset{variable}{\uparrow}}{T}\, dn = Q \tag{i}$$

Now T under the integral is a variable and changes as n changes, and maybe the ideal gas law will somehow relate and allow us to eliminate one of these quantities. So let us play with the ideal gas law

$$pV = n\,RT, \quad or \quad nT = \frac{pV}{R} = constant$$

therefore

$$n_1 T_1 = n_2 T_2, \tag{ii}$$

also

$$n = \frac{n_1 T_1}{T} \quad and \quad \underset{vessel}{dn_{in}} = \frac{n_1 T_1}{-T^2}\, dT = -dn_{out} \tag{iii}$$

These are useful results. Replace them in Eq. i to give

$$n_2 T_2 \left(\frac{\cancel{n_1 T_1}}{\cancel{n_2 T_2}} \right) c_v - \cancel{n_1 T_1} c_v + c_p \int_{T_2}^{T_1} T \left(\frac{n_1 T_1}{-T^2} \right) dT = Q$$

or

$$Q = \frac{c_p p V}{R} \, \ell n \frac{T_2}{T_1} \tag{iv}$$

Replacing values gives the heat needed

$$Q = \frac{29.10(100\ 000)\ (20\ 000)}{8.314} \, \ell n \frac{298}{273} = 613\ 372\ 920 \text{ J/day.}$$

The number of 1.5 kW heaters needed is

$$\frac{613 \times 10^6 \text{ J/day}}{24 \times 3600 \text{ s/day}} \left(\frac{\text{heater}}{1500 \text{ W}} \right) = 4.73$$

$$\text{or } \underline{5 \text{ heaters}} \quad \longleftarrow$$

Method 2. Approximate solution

Suppose we didn't catch the ideal gas trick of Eqs. ii and iii. Then we'd have to take an average value for T and solve Eq. i directly. Let's do this. Then Eq. i becomes

$$n_2 c_v T_2 - n_1 c_v T_1 + c_p \overline{T}(n_1 - n_2) = Q \tag{v}$$

Evaluating terms for air gives

$$n_1 = \frac{p_1 V_1}{R\,T_1} = \frac{(100\ 000)\ (20\ 000)}{(8.314)\ (273)} = 881\ 165 \text{ mol}$$

Similarly

$$n_2 = 807\ 242 \text{ mol}$$

$$n_1 - n_2 = 73\ 923 \text{ mol}$$

$$c_p = 29.10 \quad \text{and} \quad c_v = 29.10 - 8.314 = 20.79 \text{ J / mol·K}$$

$$\overline{T} = \left(\frac{273 + 298}{2} \right) = 285.5 \text{K.}$$

Replacing in Eq. v gives

$$(807\ 242)\ (20.79)\ (298) - (881\ 165)\ (20.79)\ (273) + (29.10)\ (286.5)(73\ 923) = Q$$

or

$$\cancel{5001 \times 10^6} - \cancel{5001 \times 10^6} + 616\ 307\ 140 = Q$$

Therefore, the number of heaters needed is

$$\frac{616 \times 10^6}{24 \times 3600} \frac{1}{1500} = 4.76$$

or <u>5 heaters</u>   ⟵

Method 3.   Clever shortcut approach

Forget all about the unsteady state energy equations of this chapter. Just recall that with $W = 0$, Eq. 8–2 gives

$$Q = \int_{273}^{298} n\, c_p dT \qquad \text{(vi)}$$

gas in the hall, n is not constant

But from ideal gas laws

$$n = \frac{pV}{RT} \qquad \text{(vii)}$$

Combining Eq. vi with vii gives

$$Q = \int_{T_1}^{T_2} \frac{pV}{RT} c_p dT = \frac{pV\, c_p}{R} \ell n \frac{T_2}{T_1}$$

This expression is identical to Eq. iv so the rest of the solution follows directly, from Method 1.

PROBLEMS

1. A rocket rises from launch pad (state 1) into a stationary orbit 37 000 km above earth (state 2). In symbols write the energy balance for this process, dropping all unnecessary terms.

2. I blow up a flaccid balloon (state 1—no air inside) to a big sphere (state 2). Write the energy balance for this process, dropping all unnecessary terms.

3. A tank-destroying rocket is shot from a soldier's shoulder. Write the energy balance for this rocket, before launching (state 1) and as it flies towards its target (state 2). Drop all unnecessary terms.

4. The Flying Scotsman, Britain's most famous train, is powered by a steam locomotive and it holds the record for the longest and fastest nonstop passenger service in Britain. It starts at London (state 1) and ends its nonstop

run in Edinburgh (state 2). Write the energy balance for the locomotive with its coal and water containing tender.

5. Hot liquid water at its boiling temperature (60°C) and its boiling pressure (19 940 Pa) enters and fills a completely evacuated and completely insulated vessel. Find the temperature of water in the vessel.

6. An insulated evacuated tank 0.1 m³ in volume is connected by a valve to a 10 bar 200°C steam line. The valve is quickly opened, the pressure in tank and line equalize, and then the valve is closed. What is the temperature in the tank?

7. At the commercial suppliers of industrial gases the standard gas tanks to be recharged are first completely evacuated to remove impurities, then refilled and returned to the customer at 136 bar. What should be the pressure in the large supply tanks to guarantee that we do receive gases at the claimed pressure? Take nitrogen, for example.

8. A steam line (0.6 MPa and 200°C) is connected to a rigid evacuated tank immersed in a constant temperature oil bath kept at 200°C. The valve between line and tank is opened, steam rushes into the tank until its pressure equals that of the line. Find the amount of heat transferred from tank to bath or the other way around . . . if any, per kg of steam entering the tank.

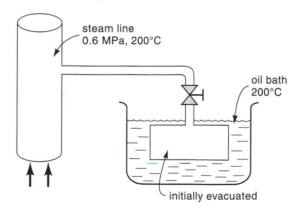

9. Gases Inc. claims that their helium contains less than 0.1% air as impurity and charges me accordingly when I come for a refill. However, I suspect that they are sharp operators, so let me try to check their claim. For this I fit my insulated evacuated tank with a thermometer; on refilling from their supply reservoir I note that the temperature in my tank reads 207°C while the filling line reads 27°C. Are the suppliers honest? If not, can you estimate the fraction of air in the helium line?

10. In Example 14–2 we opened the valve and let the pressure in the tank come to 1 bar. Suppose we didn't let the pressure equalize, but quickly shut off the air inlet valve when $p_{tank} = 0.5$ bar. Then we go off for a leisurely lunch. When we return what does the pressure gauge read?

11. A large high pressure tank contains n_1 mols of a general nonideal gas at T_1 and p_1. It develops a crack and gas slowly leaks out but the tank doesn't change temperature. After a while n_{out} moles of gas have leaked out. Develop an expression to tell how much heat loss or heat gain has occurred during this process.

12. In problem 11 the tank contained a nonideal gas. Can the expression you obtained in problem 11 be simplified if the gas is an ideal gas?

13. I have a 1 lit tank connected to a steam line. The valve is open and the tank and line are full of saturated steam at 200 kPa and 120°C. I want this tank to be full of liquid at this pressure. To do this I plunge the tank in a pool of ice and water. Steam condenses, more steam enters the tank, and eventually all in the tank is liquid at 120°C. How much ice must I melt?

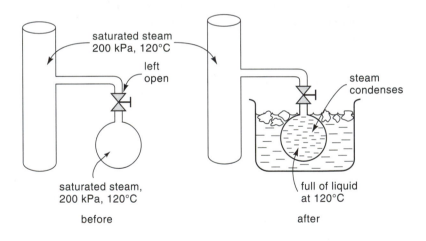

14. I have a 10-liter tank connected by an insulated tube to a steam line. The valve between line and tank is open and the tank and line are full of steam at 400°C and 5 bar. I want this tank to be full of water at 0°C. To do this I have to introduce more steam, condense and cool it, all at 5 bar. For this I plunge the tank in ice water. How much ice must I melt?

15. Steam passes through a short hose into a well-insulated barrel of water, at 1 bar, where it condenses, with the following result.

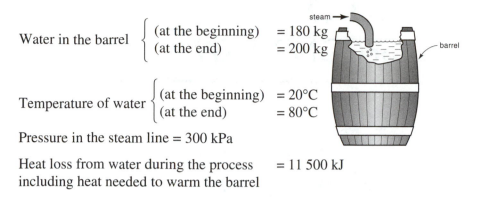

Water in the barrel $\begin{cases} \text{(at the beginning)} & = 180 \text{ kg} \\ \text{(at the end)} & = 200 \text{ kg} \end{cases}$

Temperature of water $\begin{cases} \text{(at the beginning)} & = 20°C \\ \text{(at the end)} & = 80°C \end{cases}$

Pressure in the steam line = 300 kPa

Heat loss from water during the process   = 11 500 kJ
including heat needed to warm the barrel

Find the condition of the steam (enthalpy, quality, and temperature).

16. When the pressure in our helium tank drops to 3 bar we cannot use the helium in our process. We then take our tank to the refilling station, connect it to the 30 bar line, refill rapidly, and pay the refill charge. The attendant there recommends that we first let the pressure drop to 1 bar before refilling, saying we'd get more helium that way. Please check to see how much more helium we'd get if we followed his recommendation.

CHAPTER 15

THE SECOND LAW

Experience shows that all sorts of events happen only one way in time. For example,

- Cream mixes with coffee, but once mixed it doesn't unmix.
- Books fall from table to floor, but never jump up from floor to table.
- A cup of hot water poured into a pan of cold water will give warm water. You can't scoop up a cup of hot water from this pan.
- Drop a plate and it breaks. Have you ever seen the pieces slide together to remake a good unbroken plate?
- Ponce de Leon searched for his fountain of youth. Old men are still dreaming of regaining theirs, especially when they see young ladies, but . . .
- We know that air consists of 21% O_2 and 79% all else. Have you ever read of these gases separating so that students in a classroom become high on enriched air while the professor gags, gasps, and dies an agonizing death from lack of oxygen? Students may wish it, but it will not happen. Gases will mix, but will not unmix, of themselves.

There are similar observations in other fields. For example,

- In trucking: Have you ever seen a truck with two trailers try to back down a road?
- In textbook writing: Have you ever seen a second edition of a thermo text that is less wordy than a first edition?

Science aims to explain[1] what goes on around us, and has left it to thermo to try to develop a general statement, a law, a scientific law, that somehow accounts for all these different one-way phenomena. Thermo has taken this on in what is called the *second law*. This law is stated in many different ways. Let us look at some of these.

1. *General*: Time goes in one direction, certain events are ordered. More colorfully we may say, "The second law is time's arrow." As an example, when you look at a home movie you can only tell if it is being shown forward or backward when you see an instance of the second law (Figure 15–1).

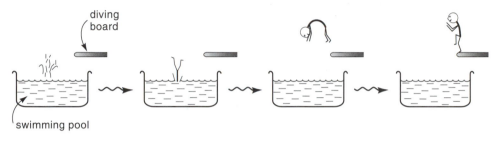

Figure 15–1.

2. *General:* Every system left to itself (isolated) changes rapidly or slowly and eventually comes to a state of rest (equilibrium); in other words, in an isolated system things always move in one direction (Figure 15–2).

---

[1]Today the word "explain" means to fit the observation into a general pattern such that we can say "you see, this observation is a special case of that theory or law." In medieval times the word "explain" would have meant to fit the observation to some theological or philosophical dogma or teaching. In that period what Aristotle said was taken to be correct and not to be questioned. For example, he said:
- Men had more teeth than women. Don't bother checking because you may get your finger bitten off.
- There are only seven wandering heavenly bodies, each circling the earth in a perfectly circular path (it had to be circular because a circle is a perfect figure, and God wouldn't make anything that wasn't perfect). When Galileo trained his first telescope on Jupiter and saw four moons going around it clerics refused to even look. It couldn't be, it had to be an illusion, you couldn't have eleven wanderers—with four daring to not circle Earth. Seven was the magic number—seven days in the week (named after the then known seven heavenly bodies—the sun, moon, Mars, Mercury, Jupiter, Venus, Saturn), seven openings into your head and soul. Also, a horse had 4×7 teeth. Again, don't bother to count.

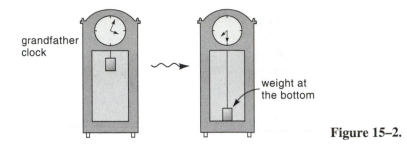

grandfather
clock

weight at
the bottom

Figure 15–2.

3. *In terms of heat flow:* It is not possible to devise a process whose sole result is to transfer heat from a low temperature object to a high temperature object (Figure 15–3); in other words, heat will not flow of its own accord "uphill."

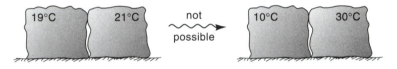

Figure 15–3.

4. *In terms of heat and work:* No apparatus can operate so that its only effect is to convert heat completely into work without also causing a change in state of some other body (Figure 15–4).

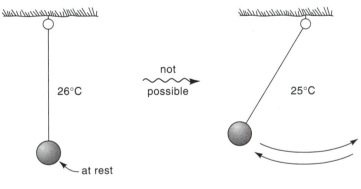

Figure 15–4.

5. *Quantitative measures:* A system left to itself moves from
  • A more ordered to a more disordered state
  • A state of lower probability to one of higher probability
  • A more recognizable to a less recognizable state
  • A state with much information to one of less information
  • A state of more precious (or useful) energy to one of less precious energy.

This class of definitions gives us a quantitative measure of direction in time—just calculate the amount of order, the probability, the amount of information (Figure 15–5). These definitions are useful in various fields—information theory, statistical mechanics, and statistical quantum thermodynamics, whatever that means.

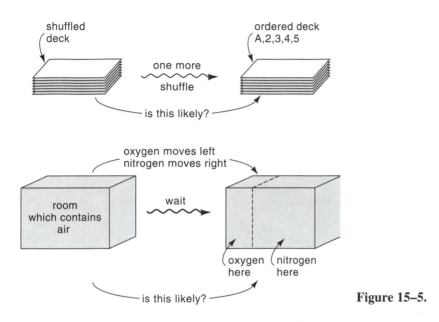

Figure 15–5.

6. *Useful thermodynamic statement:* Here we introduce a quantity called *entropy* S,[2] which allows us to determine what can happen and what cannot happen in the real world. It tells the direction of time, what state represents "before," what state represents "after." Thus, a change can take place, or is not prohibited from taking place:

in any isolated system if ... $\Delta S_{system} \geq 0$

for any system having
interaction with its $\Delta S_{system} + \Delta S_{surroundings} \geq 0$ (15–1)
surroundings if ...

*each may increase or decrease
but the sum total must not decrease*

[2]The word "entropie" was coined by Rudolph Clausius in Berlin in 1854.

What this says is that the total entropy of an isolated system will not decrease, or $\Delta S_{total} \geq 0$.

Entropy is a property of the system, as is internal energy U; and just as we usually measure ΔU for a change, so it is with entropy; we usually measure ΔS, or entropy, above some arbitrary state, say 0°C and 1 bar.

A.   MEASURING ΔS

If a system goes from state 1 to state 2, ΔU is found by measuring Q and W to and from the system and then using the first law expression $\Delta U = Q - W - \Delta E_p - \Delta E_k$. In a similar manner the entropy change in going from state 1 to state 2 is found as follows

This is the Q when all mechanical energy changes occur reversibly, without friction and where the system is uniform in T at any instant, not hot on one side, cold on the other side

$$\Delta S_{system,\ 1 \rightarrow 2} = \int_{state1}^{state\ 2} \frac{d\,Q_{rev}}{T_{system}} \qquad \left[\frac{J}{K}\right] \tag{15–2}$$

absolute temperature

For a reversible isothermal change Eq. 15–2 becomes

$$Q_{rev} = T_{syst}\,\Delta S \tag{15–3}$$

There are many paths for going from state 1 to state 2 (Figure 15–6). To determine ΔU it makes no difference which path we take because if W changes so does Q.

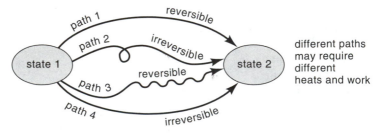

different paths
may require
different
heats and work

Figure 15–6.

For example, to get water from 20°C to 30°C

- We can just add 50 J of heat alone
- Alternatively, we can add 50 J of mechanical work by churning the water with a paddle or by adding electrical work
- We can do 70 J of work on the system and remove 20 J of heat

In all cases ΔU is the same, or

$$\Delta U = Q - W = 50 - 0 \qquad = 50$$
$$= 0 - (-50) \quad = 50$$
$$= -20 - (-70) = 50$$

However, the path *does* matter when we want to measure ΔS. The procedure is as follows:

- For the system clearly note the initial and final states and calculate the actual heat added to or removed from the system, Q_{actual}.
- If no mechanical effect intrudes such as ΔE_p, ΔE_k, expansion or contraction, friction, or other work effect, $Q_{rev} = Q_{actual}$, so use Q_{actual} directly in Eq. 15–2.
- If the mechanical energy changes involve friction, then you must devise a reversible path for the mechanical energy changes that have taken place. Calculate Q_{rev} and then ΔS for this path from initial to actual final state.

Exactly the same procedure is used to calculate $\Delta S_{surroundings}$. It may well be that one path has to be chosen for the system while a different path has to be chosen for the surroundings.

Discussion on Q_{rev} and Q_{irrev}

The distinction between these heats may be difficult to grasp. Maybe the following two illustrations will help to clarify it. First consider a 20 kg bag of sand that is dropped 50 m. If the bag is well insulated, Example 15–3 tells that it gets hotter, that it gains 9800 J of internal energy, gained from the loss of its potential energy. Since the bag is well insulated this 9800 J stays in the bag, and it gets hotter, however

$$Q_{actual} = 0$$

But in this process the lowering of the bag is done irreversibly.

If the change is done reversibly (say by raising another equivalent 50 kg weight), then we must add 9800 J of heat to the bag to increase the U of the bag to its final value, in which case

$$Q_{rev} = 9800 \text{ J}$$

We call this Q_{rev} because it refers to the situation where the potential energy change is being done reversibly.

Consider a second example, a well insulated 1 kg batch of water at 20°C, to which we add 4184 J of electrical work, represented by W_{sh}. We show this in a sketch (Figure 15–7).

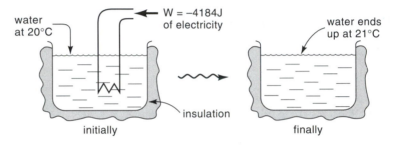

Figure 15–7.

Here $Q_{actual} = 0$. But if we wish to add the work reversibly (without generating heat), then to get the water from 20 to 21°C we must add heat. Thus

$$Q_{rev} = 4184 \text{ J}$$

Discussion about the Different Q_{rev} Values

Finally, here is something you may not be aware of. There may be any number of reversible paths from state 1 to state 2, each with its particular Q_{rev}. Although these Q_{rev} may differ from reversible path to reversible path, still ΔS is the same for all of these different paths!

You should note that S depends only on the state of the system. It is a state function (as is U, T, or H). It is not a path function as is Q or W. And the only way to calculate ΔS for a change from state 1 to state 2 is to do it with a reversible path. Section E of Chapter 16 shows that two different reversible paths, with different Q_{rev}, still have the same ΔS. The examples below illustrate these ideas that are central to thermodynamics.

EXAMPLE 15–1. **To freeze water** (Figure 15–8)

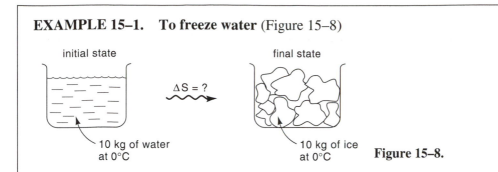

initial state   final state

ΔS = ?

10 kg of water at 0°C   10 kg of ice at 0°C   **Figure 15–8.**

Find ΔS when 10 kg of liquid water at 0°C is frozen at 0°C.

Data

For the melting of ice: $\Delta h_{s\ell} = 333$ kJ/kg (from Table 8–1)

Solution

Consider the 10 kg of water that is to be frozen to be the system and note that no mechanical energy changes are involved,[3] so apply Eq. 15–2 directly. Thus, for the system

heat is to be removed, not added so Q is negative

$$\Delta S = \int_{water}^{ice} \frac{d\,Q_{rev}}{T} = \frac{1}{T}\int d\,Q_{rev} = \frac{m(-\Delta h_{s\ell})}{T}$$

constant

$$= \frac{(10\ kg)\,(-333\ kJ/kg)}{273\ K} = -12\ 200\ \frac{J}{K}$$

EXAMPLE 15–2. **To heat water**

Find ΔS when 10 moles of liquid water go from 0°C to 100°C at 1 atm.

Data

$c_{p,water} = 4184$ J/kg·K (from Table 8–1)

---

[3]Actually, a very small amount of work (mechanical energy) was done because on freezing, water expands and pushes the atmosphere back a tiny bit—we ignore this.

Solution

Take the 10 mol H_2O to be the system. Again a negligible amount of mechanical energy is involved, so use Eq. 15–2 directly

$$\Delta S = \int \frac{d\,Q_{rev}}{T} = \int \frac{m\,c_p\,dT}{T} = m\,c_p \int_{initial}^{final} \frac{dT}{T} = m\,c_p\,\ell n \frac{T_{final}}{T_{initial}}$$

$$= (10\text{ mol}) \left(\frac{0.018\text{ kg}}{mol} \right) \left(4184\,\frac{J}{kg\cdot K} \right) \ell n \frac{373}{273} = +235\,\frac{J}{K} \quad \longleftarrow$$

EXAMPLE 15–3.   The dangers of being a beloved professor (Figure 15–9)

Just as the thermo professor was hurrying into the classroom building a big 20 kg bag of sand (c_p = 1000 J/kg·K) smashed to the ground behind him, just missing him. He looked up and saw that it fell (or dropped) 50 m from the roof deck of the building. The first, second, and third thoughts that crossed his mind were

       (a) What was ΔS of the bag of sand for this event?
       (b) What was ΔS of the surroundings?
       (c) What was ΔS_{total} for this process?

The fourth thought that came to mind was that this was a neat problem to ask his first term thermo students in their next quiz. Please answer questions (a), (b) and (c).

Additional information

Let the bag of sand be the system and let the rest of the world be the surroundings. Next, although frictional heat may have been generated when the bag hit the ground and could have heated the bag somewhat, this heat was lost to the surroundings as the bag returned to the original ambient temperature of 7°C. So take state 1 to be the bag 50 m up on the roof at 7°C, and state 2 to be the bag on the ground at its final temperature of 7°C.

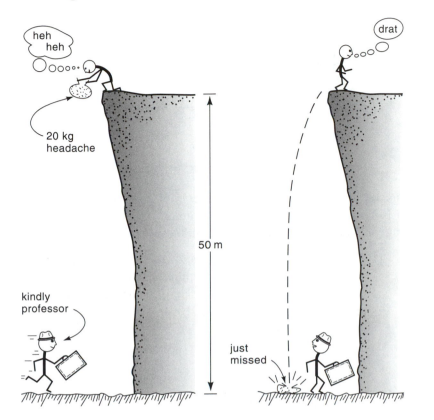

Figure 15–9.

Solution

For the system, the bag of sand, the heat actually lost to the surroundings is given by the first law, or Eq. 3–2

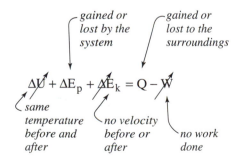

or

$$Q_{actual} = \frac{m\, g(z_2 - z_1)}{g_c}$$

$$= \frac{(20\text{kg})\,(9.8\text{ m/s}^2)\,(0 - 50\text{ m})}{(1\text{ kg·m/s}^2\text{·N})} = -9800 \text{ J}$$

lost by sand and gained by surroundings

Let us ask whether any mechanical energy effects have been involved. The answer is "yes" for the bag of sand since it has lost potential energy, but "no" for the surroundings since it has only gained heat.

(a) *For the system* (the bag of sand) the fall represented an irreversible mechanical process, so let us think up a reversible process for getting the bag from roof to ground. Here is one, just imagine a pulley system as sketched in Figure 15–10.

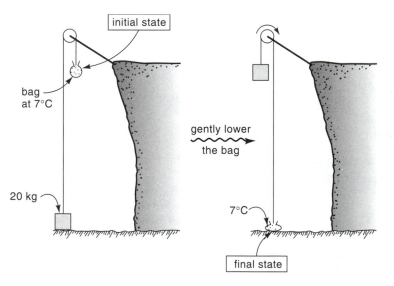

Figure 15–10.

The heat needed to get the bag to the ground reversibly is zero, hence from Eq. 15–2

$$\Delta S_{\text{bag of sand}} = \int \frac{d\,Q_{rev}}{T} = \underline{\underline{0}} \longleftarrow \text{(a)}$$

(b) *For the surroundings.* First, note that no mechanical energy changes occur, only heat effects, in that after the bag hits the ground the surroundings gain 9800 J but remain at 7°C, so

$$\Delta S_{surr} = \int \frac{d\,Q_{rev}}{T} = \frac{Q_{rev}}{T} = \frac{9800}{280}$$

positive because of heat gain

constant

$$= \frac{9800\ J}{280} = 35\ J/K \quad \longleftarrow \text{(b)}$$

(c) Overall the second law says that ΔS_{total} should be positive. This is verified since

$$\Delta S_{total} = 0 + 35 = +35\ J/K \quad \longleftarrow \text{(c)}$$

PROBLEMS

1. A 2 kg ball, rolling westward at 3 m/s collides head on with a second 1 kg ball which is at rest. The first ball leaves the collision westward at 1 m/s, the second ball also leaves westward, but at 4 m/s. The temperature is 27°C. What is the entropy change

   (a) of the first ball?

   (b) of the second ball?

   (c) overall?

2. A lump of metal sits in a bathtub and is completely covered with water. The plug is pulled and water drains from the tub. What happens to

   (a) the energy

   (b) the enthalpy and

   (c) the entropy of the lump of metal?

   The temperature remains constant throughout and the metal block is incompressible.

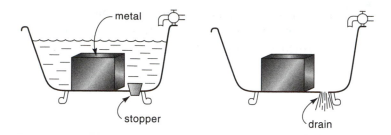

3. A weight hangs on the end of a very thin nylon thread (neglect its mass) from an insulated isolated black box. The thread is being drawn up into the box. What is happening

   (a) to the energy and

   (b) to the entropy of the box?

   Is it going up or down; is it unchanged or unknown?

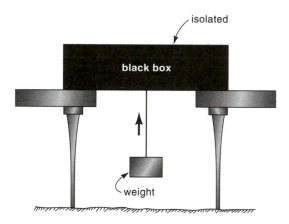

4. A 1 liter block of wood (ρ = 500 kg/m³) is held at the bottom of a pool of water. It is released and it rises to the surface 2 m above. By contact with the surroundings the temperature of the block and of the water before and after is 21°C. Find the entropy change of block, of pool, of surroundings, and overall.

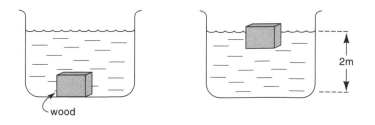

5. A 2 kg ball, rolling northward at 3 m/s collides head on with a second 1 kg ball, which is at rest. The two balls stick together and leave northward at 2 m/s. The temperature is about 27°C. What is the total entropy change involved in this collision?

6. A 10 kg iron ball is sitting by the side of a swimming pool. I carefully and gently lower it 3 m without friction to the bottom of the pool. Everything stays isothermal at 25°C.

   (a) Find the change in internal energy, total energy, and entropy of the ball.

   (b) Find the change in internal energy, total energy, and entropy of pool and water.

   (c) Find the change in total energy and entropy of the well-insulated machine that is used to lower the weight.

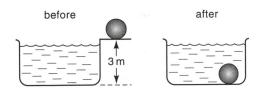

7. A pendulum 2.04 m long consists of a 5 kg mass at the end of a thread. The pendulum is held in a horizontal position on a warm day (300 K), it is released, it swings, and it eventually comes to rest at 300 K. Find

   (a) $\Delta U_{pendulum}$ **(d)** $\Delta S_{pendulum}$

   (b) $\Delta U_{surroundings}$ **(e)** $\Delta S_{surroundings}$

   (c) ΔU_{total} **(f)** ΔS_{total}

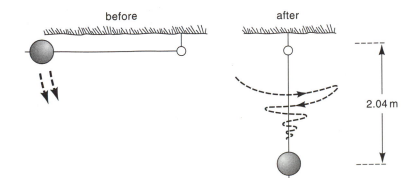

8. A block of metal (8 kg, c_p = 1 kJ/kg·K, T = 400 K) is placed into a large swimming pool filled with water (T = 300 K), and comes to equilibrium. For this process find

(a) ΔS_{metal} **(d)** ΔH_{metal}

(b) ΔS_{pool} **(e)** ΔH_{pool}

(c) ΔS_{total} **(f)** ΔH_{total}

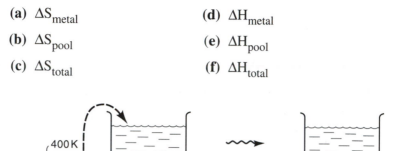

9. Below are two sketches of an isolated system. Is change A possible? Is change B possible?

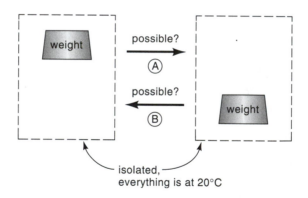

10. Below are two sketches of an isolated system. Does change A or change B violate the first law? Do either of these changes violate the second law?

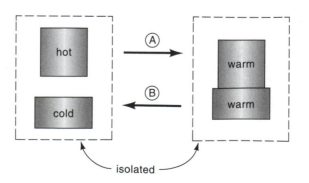

11. A rock is flung from a slingshot. Does the entropy of the rock increase, decrease, or stay constant in going from state 1 to state 2?

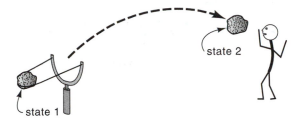

12. A hot block (1 kg, c_p = 1000 J/kg·K, 1000 K) is set upon a cold block of solid (2 kg, c_p = 1000 J/kg·K, 100 K). They reach 500 K in surroundings, which are at 300 K. Does the entropy increase, decrease, or stay unchanged

  (a) for the hot block?

  (b) for the cold block?

  (c) for the surroundings?

  (d) overall?

  (e) What does all this mean?

13. Find the ΔS for Captain Shultz's spectacular performance as described in problem 8–9.

14. Find ΔS for the return to Earth of Nepal's Annapurna space ship of problem 8–12.

15. Here is an example of the second law for 3-year-olds.

> Humpty Dumpty sat on a wall
>> Humpty Dumpty had a great fall
> All the king's horses and all the king's men
>> Couldn't put Humpty together again.

In case you don't know, Humpty was an egg and this old English nursery rhyme teaches that scrambling an egg is certainly an irreversible process. But is it? Suppose you fed the scrambled egg, shell and all, to a hen who then laid another egg, wouldn't this represent a reversible process? Your opinion, please.

CHAPTER 16

IDEAL GASES AND THE SECOND LAW

Many situations require that we find the entropy change of ideal gases. So let us develop the various formulas needed. In essence, then, we will extend the derivations of Chapter 11, and will go one step further to find the ΔS for these specific changes.

Recall from Chapter 11, for n moles of gas, ignoring ΔE_p and ΔE_k, that

$$pV = nRT \qquad \xrightarrow[\text{is constant}]{\text{assume } c_p} \qquad \Delta U = nc_v\Delta T \tag{16-1}$$
$$c_p = c_v + R \qquad \qquad \qquad \Delta H = nc_p\Delta T$$

and from Eq. 15–2 that

$$\Delta S = \int \frac{dQ_{rev}}{T}$$

where Q_{rev} refers to heat effects when the mechanical energy changes to the system all occur reversibly. If these ME effects are irreversible and cause friction and raise the internal energy of the gas, replace this process with one whose ME effects are reversible and then add the necessary heat from the surroundings. Thus we always talk about Q_{rev} and W_{rev}.

As illustration consider the following two situations that have the same initial and final states (Figure 16–1).

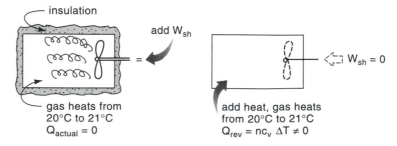

insulation

add W_{sh}

$W_{sh} = 0$

gas heats from 20°C to 21°C
$Q_{actual} = 0$

add heat, gas heats from 20°C to 21°C
$Q_{rev} = nc_v\,\Delta T \neq 0$

Figure 16–1.

The first situation has irreversible ME effects, so replace it with the second situation and use this to calculate ΔS. So to calculate ΔS for any change, reversible or irreversible, calculate Q_{rev} and W_{rev}, and use this to determine ΔS. In all the following equations c_p and c_r stay constant.

A. CONSTANT VOLUME PROCESS

From first law, at constant volume,

$$W_{rev} = 0, \quad \text{so} \quad Q = \Delta U = nc_v \Delta T$$

therefore

$$\boxed{\Delta S = \int \frac{dQ_{rev}}{dT} = \int \frac{nc_v dT}{T} = nc_v \; \ell n \frac{T_2}{T_1} = nc_v \; \ell n \frac{p_2}{p_1}}$$

and $\quad W_{rev} = 0$

(16–2)

B. CONSTANT PRESSURE PROCESS

From Chapter 11, for a reversible process at constant pressure

$$W_{rev} = \int pdV = p\Delta V = nR\Delta T$$

and

$$\Delta U = nc_v \Delta T$$

So from the first law, Eq. 3–2,

so $\quad Q_{rev} = \Delta U + W_{rev} = nc_v \Delta T + nR\Delta T = nc_p \Delta T$

and

$$\boxed{\Delta S = \int \frac{dQ_{rev}}{T} = \int \frac{nc_p dT}{T} = nc_p \; \ell n \frac{T_2}{T_1} = nc_p \; \ell n \frac{V_2}{V_1}}$$
$$W_{rev} = nR(T_2 - T_1)$$

(16–3)

C. CONSTANT TEMPERATURE PROCESS

Here $\Delta U = 0$, so from the first law

$$Q_{rev} = W_{rev} = \int p dV$$

or

$$dQ_{rev} = p dV = \frac{nRT}{V} dV$$

So the entropy change is

$$\Delta S = \int \frac{dQ_{rev}}{T} = \int \frac{nRT}{VT} dV = nR \ \ell n \frac{V_2}{V_1} = -nR \ \ell n \frac{p_2}{p_1}$$

(16–4)

and

$$W_{rev} = -nRT \ \ell n \frac{p_2}{p_1}$$

D. IN GENERAL, GOING FROM P_1 V_1 T_1 TO P_2 V_2 T_2

In terms of p_1 T_1 and p_2 T_2, consider the following two-step process (Figure 16–2).

$$
\begin{array}{ccc}
p_1\ T_1 & \xrightarrow[p_1\ =\ constant]{1^{st}\ step} & p_1\ T_2 & \xrightarrow[T_2\ =\ constant]{2^{nd}\ step} & p_2\ T_2
\end{array}
$$

Figure 16–2.

and from the expressions developed above

$$\Delta S = nc_p \ \ell n \frac{T_2}{T_1} - nR \ \ell n \frac{p_2}{p_1}$$

$$\overset{\uparrow}{\text{1st step}} \qquad \overset{\uparrow}{\text{2nd step}}$$

(16–5)

Similarly, in terms V_1 T_1 and V_2 T_2, we find from a two step process

$$\boxed{\Delta S = nc_v \; \ell n \frac{T_2}{T_1} + nR \; \ell n \frac{V_2}{V_1}} \qquad (16\text{--}6)$$

and, in terms of $p_1 \, V_1$ and $p_2 \, V_2$

$$\boxed{\Delta S = nc_v \; \ell n \frac{p_2}{p_1} + nc_p \; \ell n \frac{V_2}{V_1}} \qquad (16\text{--}7)$$

The above expressions tell that if you know the initial and final conditions plus c_p then you are able to evaluate

$$\Delta U, \Delta H \text{ and } \Delta S$$

for *any* change (any path taken) of an ideal gas. Please be reminded that ΔU, ΔH, and ΔS are state functions (depend only on initial and final states, not on the path taken in moving from one state to another), since they do not include Q and W in their formulas.

E.   REVERSIBLE WORK

Consider the reversible work involved in going from $p_1 T_1$ to $p_2 T_2$. One can take various paths. Two are shown in Figure 16–3.

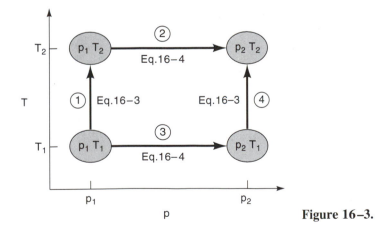

Figure 16–3.

From the equations just developed we have, for paths 1+2,

$$W_{1+2} = nR(T_2 - T_1) - nRT_2 \; \ell n \frac{p_2}{p_1}$$

and for paths 3 + 4

$$W_{3+4} = -nRT_1 \, \ell n \frac{p_2}{p_1} + nR(T_2 - T_1)$$

Comparing these expressions shows that the work required differs for these paths. This is an example of the generalization that in going from p_1T_1 to p_2T_2,

---

- The work, whether reversible or not, is always path dependent.
- Therefore, the heat involved also is path dependent.

- However, $\dfrac{Q_{rev}}{T}$ or $\displaystyle\int \dfrac{dQ_{rev}}{T}$ —in other words ΔS—is independent of the path taken. (Verify this by examining Eq. 16–5, 16–6, or 16–7).

F. ADIABATIC REVERSIBLE PROCESSES (Q = 0; ΔS = 0)

This represents the ideal machine. For example, consider a rotating compressor or turbine (Figure 16–4).

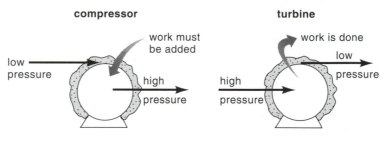

Figure 16–4.

Without friction or heat interchange with the surroundings the turbine and compressor behave just the reverse of each other. Let us now see what happens to the fluid passing through an adiabatic reversible machine. For each mol flowing

From First law (adiabatic)	$q_{actual} = 0$
and for flow systems	$-w_{sh} = \Delta h + \Delta e_p + \Delta e_k$
From Second law (reversible adiabatic)	$\Delta s_{fluid} = \displaystyle\int \frac{dq_{rev}}{T} = 0$
and with Eq. 11–15 (reversible adiabatic)	$\dfrac{T_2}{T_1} = \left(\dfrac{p_2}{p_1} \right)^{\frac{(k-1)}{k}}$

(16–8)

These machines are considered ideal because they require less work to compress and produce more work to expand than most real machines with their frictional effects. Example 16–2 illustrates the analysis of these machines.

For an adiabatic but irreversible process (heat generation by friction and added to the fluid) we have

$$\Delta S_{\text{fluid}} > 0 \tag{16-9}$$

For a nonadiabatic irreversible process (heat can be added or removed from the fluid as it passes through the machine)

$$\Delta S_{\text{fluid}} \text{ can be } >, <, \text{ or } = 0 \tag{16-10}$$

EXAMPLE 16–1.   Expansion of a gas (Figure 16–5)

The left half of an insulated constant volume container contains air, and is separated from the evacuated right half by a diaphragm. The diaphragm ruptures and air spreads throughout. Find ΔS for this process. The sketch below gives numerical quantities.

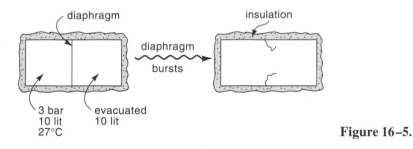

Figure 16–5.

Solution

First, find the final state plus the heat and work actually involved from the first law. Then worry about the entropy change. So from the first law for the whole isolated system of constant volume

$$Q = 0 \text{ and } W = 0$$

Therefore ΔU = 0. So for the ideal gas

$$T_{\text{final}} = 300 \text{ K}, \ V_{\text{final}} = 20 \text{ lit}, \ p_{\text{final}} = 1.5 \text{ bar}$$

and

$$n = \frac{pV}{RT} = \frac{(150\ 000)(0.020)}{(8.314)(300)} = 1.2 \text{ mol}$$

Since this process involves an explosion, turbulence, and friction—hence mechanical effects—and since it is irreversible, to determine Δs we must think up a reversible process to go from the same initial to the same final state. Any one of a number of processes would do; however, the simplest is an isothermal reversible expansion from initial to final state. For this Eq. 16–4 gives

$$Q_{rev} = W_{rev} = nRT \, \ell n \frac{V_{final}}{V_{initial}}$$

$$\Delta S = \int \frac{dQ_{rev}}{T} = nR \, \ell n \frac{V_{final}}{V_{initial}}$$

$$= (1.2 \text{ mol}) \left(8.314 \, \frac{J}{mol \cdot K} \right) \ell n \frac{20}{10} = 6.92 \, \frac{J}{K} \longleftarrow$$

the "+" for an isolated system means that this change represents a move towards equilibrium. The reverse, all the gas collecting on the left could not happen of itself.

Note. $Q_{actual} = 0$ because the process is adiabatic. However, this was an irreversible process, we had to devise a reversible path and use the Q_{rev} to calculate ΔS. And as shown above $Q_{rev} \neq 0$.

EXAMPLE 16–2.   Making money from wasted air (Figure 16–6)

Presently, high-pressure air (n = 20 lit/s, T = 300 K, p = 10 atm) is vented to one atmosphere without getting anything useful from it. We are considering installing a turbine with an electricity generator to recover some of the available energy presently being lost. Find the ideal power generated for adiabatic reversible operations of the turbine, and the money recovered per 30 day month, if energy is worth 7¢/kW•hr.

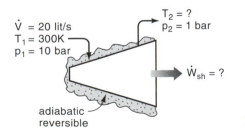

Figure 16–6.

Solution

First, find the molar flow rate

$$\dot{n} = \frac{p\dot{V}}{RT} = \frac{(1013\ 250)(0.020)}{(8.314)(300)} = 8.125\ \frac{\text{mol}}{\text{s}}$$

To evaluate the work use Eq. 8. Thus for adiabatic reversible operations

$$T_2 = T_1 \left(\frac{p_2}{p_1}\right)^{\frac{k-1}{k}} = 300\left(\frac{1}{10}\right)^{\frac{0.4}{1.4}} = 155\ \text{K}$$

$$\dot{W}_{rev} = -\Delta\dot{H} = -\dot{n}c_p(T_2 - T_1)$$

$$= -(8.125)\ (29.10)\ (155 - 300) = 34\ 283\text{W}$$

Money recovered

$$= \left(34\ 283\frac{\text{J}}{\text{s}}\right)\left(\frac{3600 \times 24 \times 30\ \text{s}}{\text{month}}\right)\left(\frac{0.2778\ \text{kW·hr}}{10^6\ \text{J}}\right)\left(\frac{\$0.07}{\text{kW·hr}}\right)$$

$$= \underline{\$1728/\text{month}} \quad \longleftarrow$$

Comment. This represents a considerable saving. Note that the exit air is very cold, 155 K, and this leads us to suspect that we somehow could recover energy from it. This is so. So consider a reversible isothermal expansion of the original gas as sketched in Figure 16–7.

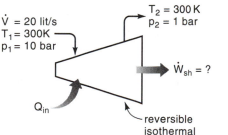

Figure 16–7.

From Eq. 16–4

$$\dot{W}_{sh} = -\dot{n}RT\ \ell n\frac{p_2}{p_1} = -(8.125)\ (8.314)\ (300)\ \ell n\frac{1}{10} = 46\ 662\ \text{W}$$

Comparing the two results we see that the isothermal reversible expansion gives 46.6/34.3 = 1.36, or 36% more power than the adiabatic reversible expansion.

EXAMPLE 16–3.   Painless surgery (Figure 16–8)

A salesman for Slice and Dice, Inc., offered our surgical team a new type of scalpel, one that he claims cuts absolutely painlessly. It works by blowing very, very cold air onto the tissue to be cut. Tissue freezes, feeling disappears, and there is no need for anesthetics. Fabulous!!

He says that the heart of this device is a Hilsch tube, a device that splits a stream of high-pressure air (p_1 = 1.5 bar, T_1 = 27°C) into two equimolar streams, one hot and the other cold, both at lower pressure ($p_2 = p_3$ = 1 bar). The cold stream is then blown at the scalpel blade.

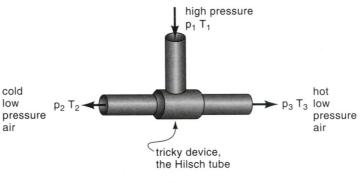

Figure 16–8.

The salesman claims that the cold air is at −123°C. I don't believe that the air could get that cold with so simple a device. Would you please determine whether his claim violates the laws of thermo.

Solution

For any real process occurring in an isolated flow stream the second law says that $\Delta S_{total} \geq 0$. Let's see whether the air passing through the Hilsch tube satisfies this requirement. Also assume adiabatic reversible operation of the Hilsch tube because this will give the coldest temperature.

Basis. Consider two moles of entering air. Then from the first law for an isolated flow stream we have

$$n_1 h_1 = n_2 h_2 + n_3 h_3 \ldots \text{where} \quad \begin{cases} n_1 = 2 \\ n_2 = 1 \\ n_3 = 1 \end{cases}$$

or

$$n_1 c_p T_1 = n_2 c_p T_2 + n_3 c_p T_3$$

If the salesman's claim is correct then

$$2(300) = 1(150) + 1(T_3)$$

or

$$T_3 = 450 \text{ K}$$

From the second law for two moles of entering ideal gas, Eq. 16–5 gives

$$\Delta S_{total} = \Delta S_{cold\ side} + \Delta S_{hot\ side}$$

$$= \left[n_2 c_p\ \ell n \frac{T_2}{T_1} - n_2 R\ \ell n \frac{p_2}{p_1} \right] + \left[n_3 c_p\ \ell n \frac{T_3}{T_1} - n_3 R\ \ell n \frac{p_3}{p_1} \right]$$

Replacing values gives

$$\Delta S_{total} = \left[29.10\ \ell n \frac{150}{300} - 8.314\ \ell n \frac{1}{1.5} \right] + \left[29.10\ \ell n \frac{450}{300} - 8.314\ \ell n \frac{1}{1.5} \right]$$

$$= [-16.80] + [15.17] = -1.63 \frac{J}{2\ mol \cdot K}$$

Since $\Delta S_{total} < 0$ this device cannot get the air as cold as claimed by the salesman.

PROBLEMS

1. An insulated tank (V = 8.314 lit) is divided into two equal parts by a thin membrane. On the left is an ideal gas at 1 MPa and 500 K, on the right is a vacuum. The membrane ruptures with a loud bang.

    (a) How many moles of gas are in the tank?

    (b) What is the final temperature in the tank?

    (c) What is ΔS_{total} for this process?

2. An ideal gas (c_p = 30 J/mol·K) leaks out of a large tank through a slightly cracked insulated valve to a 100 kPa atmosphere. When the gas in the tank is at 300 K and 1 MPa,

    (a) Find the temperature of the gas just downstream from the valve.

    (b) Find Δh across the valve.

    (c) Find Δu across the valve.

    (d) Find Δs across the valve.

3. Air at 100 kPa and 300 K passes through a battered, squeaky old compressor that I think needs oiling. However, it is well insulated. The air leaves at 500 kPa and 600 K. This is too hot so we next pass the air through a heat exchanger where the air cools to 300 K and the pressure goes down to 450 kPa.

    (a) What is the entropy change per mole of air for this change from point 1 to point 2, from point 2 to point 3, and the overall change, thus going from point 1 to point 3?

    (b) What is the work needed/mol for this compression (step 1–2)?

    (c) What would be the work needed for a brand new and efficient adiabatic reversible compressor compressing the air to 500 kPa?

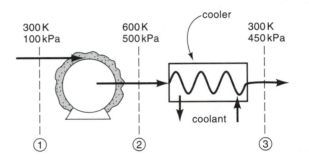

4. I'm pooped. On this warm 300 K day I've blown that giant balloon to 1 m^3 in volume and 500 kPa pressure, and it took so long that the balloon remained at room temperature. Guess what—I plan to pop the balloon. It will be spectacular and will make me completely deaf. Never mind that, please evaluate,

    (a) ΔH for the balloon's gas, from before to immediately after the pop.

    (b) ΔS for the balloon's gas, from before to immediately after the pop.

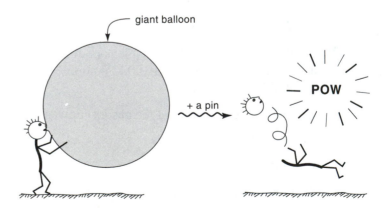

5. Referring to the previous problem where I pop a balloon, find the entropy change of the balloon gas from before the pop to a long enough time after the pop so that the temperature of the gas returns to 300 K.

6. Referring to Example 3, what air inlet pressure to an ideal Hilsch tube would split an inlet gas at 300 K into equimolar outlet streams at 150 K and 450 K, both at 1 bar?

7. Air (c_p = 29.10 J/mol·K) enters a well insulated Hilsch tube at 4 bar and 27°C. This gas is split 50-50, half leaving in the hot arm, half in the cold arm of the tube, both streams at one bar. How cold can the cold gas get if the Hilsch tube is as efficient as possible?

8. A stream of air at 300K and 1 bar (state 1) is compressed adiabatically and reversibly to 6 bar (state 2), throttled adiabatically back to 5 bar (state 3), cooled in a heat exchanger to 300K, but still at 5 bar (state 4), and then passes to a reactor. Per mole of air flowing through the system calculate

$$T_2, T_3, \Delta S_{12}, \Delta S_{23}, \Delta S_{34}, \text{ and } \Delta S_{14}.$$

Did you find that $\Delta S_{14} = \Delta S_{12} + \Delta S_{23} + \Delta S_{34}$? You should have.

CHAPTER 17

ENTROPY OF ENGINEERING FLUIDS

Refrigerators, freezers, steam engines, home air conditioners, heat pumps, jet engines, auto air conditioners, dehumidifiers—these are all machines that transform heat into work, or use work to pump heat from a lower to a higher temperature. All this is done by picking the right fluid and then cleverly heating, cooling, condensing, vaporizing, compressing, and expanding it. What we wish to point out in this chapter is that to properly design and analyze the behavior and efficiency of such machines requires knowing the entropy changes involved in the working fluid of these machines.

For all ideal gases we have nice simple equations to tell us what we need to know—for example, how well a machine could work (see Example 16–2). Unfortunately, for other materials, and also when condensation and boiling occur, we have no neat generalizations to fall back upon. Each material has to be treated separately.

This chapter tells what the entropy changes of these useful fluids are, shows this graphically and in tables, and finally gives some examples to show how this information is used.

A. ENTROPY OF PURE SUBSTANCES

Knowing the pVT relationship of a fluid or solid as well as its specific heat is all that is needed to evaluate its entropy and change in entropy at different conditions. This is obtained from the basic definition of entropy.

$$\Delta S = \int \frac{dQ_{rev}}{T}, \left[\frac{J}{K}\right]$$

$$= \int_{T_1}^{T_2} \frac{mc_p dT}{T}$$

...for a change in temperature but without phase change, and at constant pressure (17–1)

$$= \frac{m\Delta h_{\ell g}}{T}$$

...for a phase change at T

Arbitrarily picking a standard state for the measurement of enthalpy (or internal energy) and for entropy, tables have been prepared for all useful engineering materials; in particular for water for high-temperature operations (steam engine, coal-fired electrical power plants), and HFC-134a, which looks like it will be the coming material for refrigeration machines. However, beware in using these tables, because different standard states are sometimes chosen. So the numbers in one table may not match the numbers in another table. This means that you may get into trouble if you use two different tables to solve a specific problem. For example, all enthalpy values for HFC-134a given in the ICI tables are 100 kJ/kg less than the values given in the duPont tables because different standard states were chosen for these tables.

Charts have also been prepared to show the change in entropy with the change in other variables. These charts are useful for analyzing the behavior of all sorts of machines, and are of particular interest to engine designers. Here we just show what these charts look like.

First, we have the Mollier diagram, the h versus s chart sketched in Figure 17–1. Accurate Mollier diagrams are not shown in this text. Note that when you boil a liquid or go from liquid to vapor you move from point A to point B in Figure 17–1.

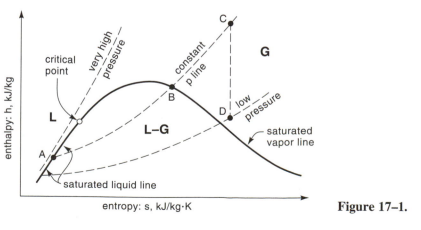

Figure 17–1.

This chart is particularly useful in evaluating the work involved in an adiabatic reversible step of an engine because

from first law	$\Delta h = w$	... for the adiabatic condition
from second law	$\Delta s = 0$	... for the reversible condition

Hence, a vertical line CD on this chart gives the work done, simply and directly, for a machine.

The T-s chart is the second useful thermo diagram. Its main features are sketched in Figure 17–2. Note that points A and B represent the transition from boiling liquid to saturated vapor. If you have a mixture of liquid and vapor, then you must interpolate between A and B. In general, per kg of mixture

$$s_{mixture} = s_g x_g + s_\ell (1 - x_g)$$

$\underset{\underset{quality}{mass\ fraction\ of\ vapor,}}{\Big\uparrow}$ (17–2)

Also, points C to D represent the drop in temperature when an adiabatic-reversible machine operates between the two corresponding pressures.

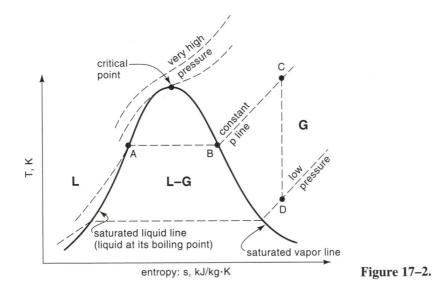

Figure 17–2.

These two charts usually ignore the lower temperature regions where liquid solidifies. This is not a useful condition since solids are not that easy to pump around in a machine. These charts are useful for quick approximate analysis of machines that transform heat to work, and work to heat. We do not present the accurate charts for water and HFC-134a in this book. This is the special interest of mechanical engineering design where giant charts are actually used.

B. GIBBS' PHASE RULE

One of the most curious and famous relations in thermo is the phase rule developed by Gibbs. In any given system, single or multicomponent, single phase or multiphase, this phase rule tells how many intensive variables you have to know to completely define the system. This number is called the degrees of freedom $\overline{df}$. For a nonreacting system, Gibbs' phase rule tells that

$$\overline{df} = C + 2 - P \qquad (17\text{–}3)$$

where   C = number of components

                                       P = number of phases

Here are some examples for a single pure material (C = 1).

- For a gas alone, or liquid alone (P = 1), Eq. 17–1 gives

$$\overline{df} = 2$$

This means that if you know p and T, or p and h, or s and h, or any pair of properties, all else is defined and can be found in the tables.

- For a two-phase (gas-liquid) mixture at equilibrium (P = 2), Eq. 17–3 gives

$$\overline{df} = 1$$

So if you know T, or p, or h, or u, or s, thus any one property, then all else is defined. Note, you cannot arbitrarily choose more than one property.

- For the triple point where vapor-liquid-solid are at equilibrium (P = 3), Eq. 17–3 gives

$$\overline{df} = 0$$

This means that you have no freedom to choose any temperature or pressure. The triple point only exists at one specific condition.

C. SIMPLE APPLICATIONS OF ENTROPY

Here are some examples of the use of the concepts of entropy and the phase rule. As you may note, these deal primarily with machines that you are trying to operate adiabatically and reversibly, hence ideally. Please remember that $\Delta s = 0$ for all adiabatic reversible processes.

We give no examples dealing with ideal gases, only examples dealing with fluids for which entropy tables are available. For ideal gases no tables are needed—just use the general equations developed in the previous chapter.

EXAMPLE 17–1.   Compression of a gas stream (Figure 17–3)

Find the work needed to compress adiabatically and reversibly to 500 kPa a stream of saturated HFC-134a gas at –40°C.

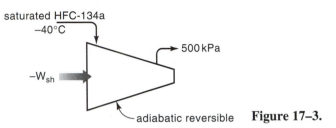

Figure 17–3.

Solution

From first law for a flow system, with negligible Δe_k and Δe_p, the work of compression needed

$$-w_{sh} = (h_2 - h_1) - \overset{=\ 0,\ adiabatic}{\cancel{q}}$$

Since the incoming fluid is saturated, phase rule says that $\overline{df} = 1$, and since we know $T_1 = -40°C$, we can find all the thermodynamic properties of the incoming fluid from the saturated table. This gives

$$p_1 = 51.14 \text{ kPa} \qquad v_1 = 0.3614 \text{ m}^3/\text{kg}$$

$$h_1 = 374.3 \text{ kJ/kg} \qquad s_1 = 1.7655 \text{ kJ/kg·K}$$

For the outgoing gas we are only given one property, its pressure. This is not enough information to evaluate its other properties. Phase rule says that we need to know two properties, so we need one more piece of information. We get this information from second law. For an adiabatic (q = 0) reversible compression we write

$$\Delta s = \int \frac{dq_{rev}}{T} = 0 \quad \dots \text{ or } \quad s_2 = s_1$$

This is just what we need. So hunting in the tables for conditions where

$$p_2 = 500 \text{ kPa and } s_2 = s_1 = 1.7655 \text{ kJ/kg·K}$$

we find $T_2 = 30°C, \qquad v_2 = 0.044\,34 \text{ m}^3/\text{kg}, \qquad h_2 = 421.3 \text{ kJ/kg}$

Therefore, the work of compression is

$$-w_{sh} = h_2 - h_1 = 421.3 - 374.3$$

$$= 421.3 - 374.3 = \underline{\underline{47 \text{ kJ/kg}}} \quad \longleftarrow$$

EXAMPLE 17–2. Adiabatic reversible turbine (Figure 17–4)

Steam at 1 MPa and 600°C enters an adiabatic reversible turbine at low velocity and exits at 150 kPa and 200 m/s. Find the work done.

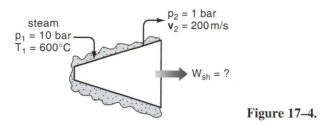

Figure 17–4.

Solution

From first law with negligible Δe_p we write for the flowing fluid

$$h_2 + \cancel{\frac{v_2^2}{2g_c}} - h_1 = \overset{= \, 0, \, adiabatic}{\cancel{q}} - w_{sh} \qquad (17\text{–}4)$$

Consider the h_1 term: Since T_1 and p_1 are known the tables give us

$$h_1 = 3697.9 \text{ kJ/kg} \qquad s_1 = 8.0290 \text{ kJ/kg·K}$$

Now consider the h_2 term: Since the process is adiabatic reversible $s_2 = s_1$. So with $s_2 = 8.0290$ kJ/kg·K and $p_2 = 100$ kPa known, hunting in the tables gives

$$T_2 = 250°C, \qquad v_2 = 2.406 \text{ m}^3/\text{kg}, \qquad h_2 = 2974.3 \text{ kJ/kg}$$

Replacing all values in Eq. 17–4 gives the work done

$$w_{sh} = h_1 - h_2 - \frac{v_2^2}{2g_c}$$

$$= 3697\,900 - 2974\,300 - \frac{(200)^2}{2(1)}$$

$$= 703\,600 \text{ J/kg} = \underline{703.6 \text{ kJ/kg}} \quad \longleftarrow$$

PROBLEMS

1. Examine the tables at the back of this book and try to find out the selected standard states for energy and entropy

   (a) for water

   (b) for HFC-134a

2. Find the properties of H_2O

   (a) at 1 bar and 20°C

   (b) at 250°C and at h = 2950 kJ/kg

   (c) liquid-vapor mixture at 2 bar and 200°C

3. Repeat Example 17–1 with but one change: The HFC-134a is being compressed in a cylinder, and is not part of a flow stream.

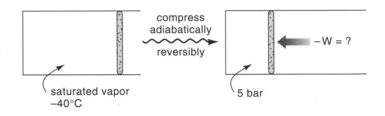

4. Repeat Example 17–2 with just one change: The incoming stream is now at 250°C, not 600°C.

5. An old flow nozzle, not known whether adiabatic or reversible, is fed steam at 300°C and 3 bar at 10 m/s and discharges steam at 250°C and 2 bar. What is the velocity of the leaving steam?

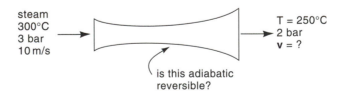

6. A large steam turbine takes in 5 kg/s of steam at 600°C and 12 bar, rejects it at 0.5 bar, and is supposed to produce $\dot{W}$ = 4MW. This seems to be a rather large power output. Is it possible, and if so, what is its efficiency compared to an adiabatic reversible turbine?

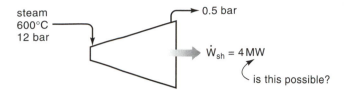

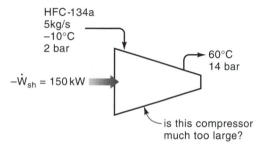

7. Looking over the plans for the HFC-134a plant something bothers me. It's that compressor shown below. From my practical experience it seems much much larger than needed. Have you somehow added a zero or two on the work term? Please check its efficiency compared with an adiabatic reversible compressor.

CHAPTER **18**

WORK FROM HEAT

In the late 1700s man discovered how to extract work from the flow of heat. The steam engine was invented. In it coal and wood were burned to generate heat that turned water into steam, which then produced work. The steam engine pumped water from mines, powered trains, and led in turn to factories, railways, steamships, mass transportation, and our modern industrial world. Just as we are at the early stage of an information revolution, with our PCs, faxes, and so on, this harness of heat to produce work was the foundation of the Industrial Revolution of the 1800s.

A nagging problem of those times was to figure out the maximum amount of work that could be obtained from a given amount of fuel. If a tender full of coal could get your train from here to there, could you make a round trip with a better steam engine? Or maybe even ten round trips if you were really clever?

Carnot's genius. Sadi Carnot, a young French military engineer, wrestled with this problem of finding the maximum efficiency of a heat engine, and by brilliant deductive reasoning, but no experimenting or tinkering with real steam engines, cracked this problem, as probably only a Frenchman could have. If only Alfred Nobel had lived before Carnot, then I'd vote a Nobel Prize for this young man.

Carnot did his analysis and gave a mathematical representation of the second law in 1811, long before the first law was clarified and properly expressed (about 1840 to 1850), so shouldn't the second law be called the first and vice versa? It probably should, but it just isn't. I don't know why.

At the time that Carnot thought and wrote about these matters, heat was considered to be a fluid, called caloric. Thus, a hot object contained lots of caloric, and removing caloric from a material cooled it. Although we replaced this concept of hot and cold with the kinetic theory of heat, which says that hot molecules move about faster than cold molecules, we still talk of heat "flowing" from hot to cold objects. Our language today still retains much of the caloric theory.

We should also mention that Carnot's analysis led to the concept of entropy, and it is still the nicest way to show its relationship with the second law, though in a somewhat restricted application. Let us start by taking the term "heat engine"[1] to mean any gadget or device for transforming heat into work, and let us consider only three kinds of operations of the engine.

- the absorption of heat from a hot constant temperature reservoir, at T_1
- the removal of heat to a cold temperature sink at T_2
- doing work or receiving work

In graphical representation we show these operations in Figure 18–1.

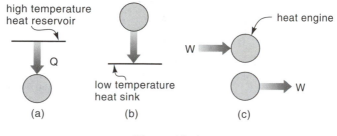

Figure 18–1.

In all other chapters of this book the signs on Q and W tell whether heat and work are added or removed from the system. In this chapter alone we deviate from this nomenclature, using |Q| and |W| to represent absolute values for heat or work, with arrows in the figures to show whether Qs and Ws are being added or removed from reservoirs or from engines.

For *continuous operations* (steady state) with *just one heat reservoir* at T there are only four conceivable combinations of operations as shown in Figure 18–2.

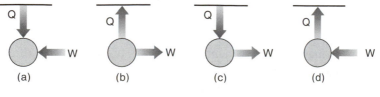

Figure 18–2.

[1]And this is where the word engineer comes from, though today the word has a broader meaning.

Now at steady state (continuous operations) schemes (a) and (b) cannot operate (violation of the first law), scheme (c) is rejected (violation of the second law), while there is no objection to scheme (d). Home plug-in electrical heaters and electrical hot water heaters all follow schedule (d)—they introduce electrical work and get heat from the engine.

This is not a very interesting scheme, but now when we consider *continuous operations* with *two heat reservoirs*, a hot one at T_1 and a cold one at T_2, then we hit the jackpot. As shown in Figure 18–3, we have eight combinations arranged as four pairs of opposite operations.

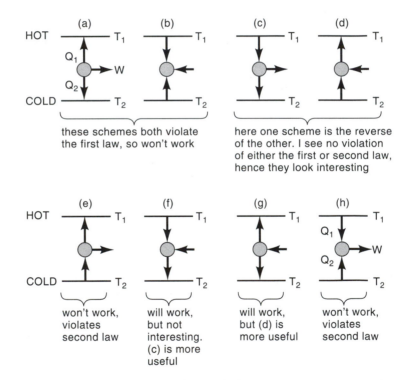

Figure 18–3.

As may be seen, schemes (c) and (d) are the reverse of each other and both can work. No other pair will do this, hence the term reversible heat engine.

- Scheme (c) is called the *Carnot heat engine* (produces work from the flow of heat).

- Scheme (d) is called the *Carnot heat pump* (pumps heat to a higher temperature by absorbing work).

A. THE CARNOT HEAT ENGINE

The above analysis shows that we cannot get work from one source of heat during steady state operation of an engine. We always need a supply of high-temperature heat and a lower temperature sink to absorb waste heat. This is shown in Figure 18–4.

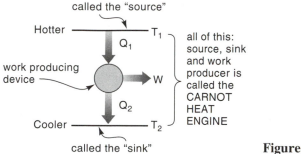

Figure 18–4.

Thus

$$|W| = |Q_1| - |Q_2|$$

Hence, the efficiency that measures the fraction of high-temperature heat that we are able to transform into work is given by

$$\eta = \frac{|W|}{|Q_1|} = \frac{|Q_1| - |Q_2|}{|Q_1|} = 1 - \frac{|Q_2|}{|Q_1|} \qquad (18\text{--}1)$$

At this point you may ask "What's so special about these two schemes being reversible and not violating either the first or second law?" Let's see what jewels of pure logic can be extracted from this.

Theorem 1. First of all, Carnot proved that *all reversible heat engines operating between the same two temperatures, T_1 and T_2, must have the same efficiency.*

Proof. Let us assume the contradictory, that two reversible engines have different efficiencies, for example, Figure 18–5.

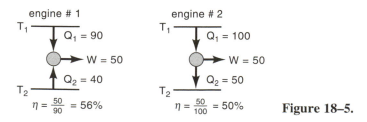

Figure 18–5.

Now, since they are reversible, reverse the second of these and interconnect the two engines (Figure 18–6).

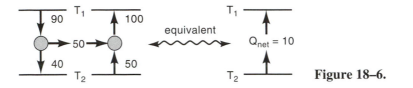

Figure 18–6.

The net result is to have heat flow uphill from the colder to the hotter reservoir without any machine or work required. Since this violates the second law, it means that the assumption does not hold, consequently Theorem 1 is proved.

> *Note.*   This kind of mathematical proof is called a "reductio ad absurdum." It means assuming the opposite of what you are trying to prove, and then showing that this leads to a contradiction. Some of the very simple and basic theorems in mathematics can only be proved this way, for example, "One cannot write the number $\sqrt{2}$ as a fraction of two numbers."

Theorem 2.   *Reversible heat engines have the highest efficiency between any two temperatures.*

> **Proof.**   Follow the same strategy as in Theorem 1. Assume that the irreversible engine has the higher efficiency, as shown in Figure 18–7.

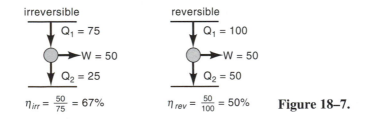

Figure 18–7.

Again, reverse the reversible engine and connect to give Figure 18–8.

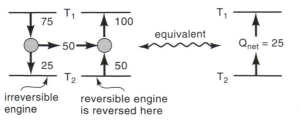

Figure 18–8.

Heat will not flow of itself uphill, so our original assumption is rejected. So comparing efficiencies we conclude that

$$\eta_{\text{reversible}} \geq \eta_{\text{irreversible}} \qquad (18\text{--}2)$$

Theorem 3. *For the same high temperature T_1, the engine that has the larger ΔT has the higher efficiency and produces more work.*

Proof. Can you reason this out from Figure 18–9? Try it.

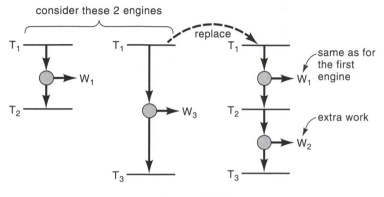

Figure 18–9.

These three theorems made Sadi Carnot immortal in science.

How to measure the temperature. So far in this chapter we haven't considered how to measure the numerical value of temperature. All we have considered are *higher*, *same*, and *lower* temperatures. But we know from the sketch above that as T_3 goes down |W| increases and |Q_2| decreases. This is the clue that led William Kelvin to join Sadi Carnot in immortality.

B. THE KELVIN TEMPERATURE SCALE

A generation after Carnot's time, Kelvin, who didn't have television to distract him, speculated that he could derive a temperature scale—let's call it T′—not based on the expansion of gases or liquids, but based on Carnot's heat engines. He asked you to consider a series of Carnot engines, each producing the same amount of work, say 10 units, as shown below, and operating at temperatures as shown in Figure 18–10.

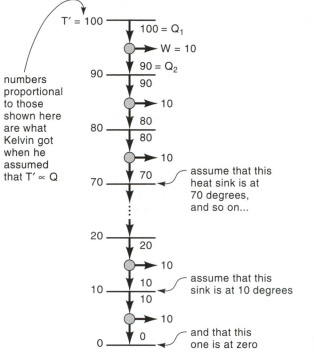

Figure 18–10.

Then he said, "Let us pick a temperature scale T′ such that T′ is proportional to Q." For example, when Q is 90, he said, "Let's say that that particular heat sink is at T = 90 degrees." Thus, for each heat engine

$$T_1' \propto Q_1 \quad \text{and} \quad \frac{T_2'}{T_1'} = \frac{|Q_2|}{|Q_1|}$$

and when combined with Eq. 18–1, gives

$$\frac{|W|}{|Q_1|} = \frac{|Q_1| - |Q_2|}{|Q_1|} = 1 - \frac{|Q_2|}{|Q_1|} = 1 - \frac{T_2'}{T_1'} = \frac{T_1' - T_2'}{T_1'}$$

or   (18–3)

$$\frac{|W|}{|Q_2|} = \frac{T_1' - T_2'}{T_2'} \quad \text{or} \quad \frac{|Q_2|}{|Q_1|} = \frac{T_2'}{T_1'}$$

This temperature scale is measured in terms of work and heat in a Carnot engine, and is known as the *Kelvin work scale*, T′. We show it in Figure 18–10. Remember that this temperature scale is something defined and derived straight from thermodynamics. There is no good reason to suspect that it has anything to do with our arbitrary scales—Fahrenheit, Celsius, Raumé, and so on, or our ideal gas scale.

But read on and let's see what surprises await us.

Let us take one mole of an ideal gas, put it in a cylinder with piston and operate it in a four-step cycle as a Carnot engine. These steps are

Step AB: Add heat at T_1 and let the one mole of gas expand while doing work.

Step BC: Expand the gas some more, this time adiabatically and reversibly, doing more work until its temperature drops to T_2.

Step CD: Remove heat at T_2 and let the volume decrease.

Step DA: Compress adiabatically until the temperature rises to T_1.

Figure 18–11 shows these 4 steps on a p-v diagram and on a T-s diagram.

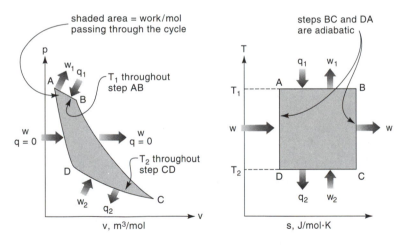

Figure 18–11.

Now from the ideal gas equations in Chapter 11, we have

$$\text{Step AB (isothermal):} \qquad |w_1| = |q_1| = RT_1 \, \ell n \frac{p_A}{p_B} \qquad (18\text{–}4)$$

$$\text{Step BC (adiabatic reversible):} \quad |w| = c_v (T_2 - T_1) \qquad (18\text{–}5)$$

$$\text{Step CD (isothermal):} \qquad |w_2| = |q_2| = RT_2 \, \ell n \frac{p_C}{p_D} \qquad (18\text{–}6)$$

$$\text{Step DA (adiabatic reversible):} \quad |w| = c_v (T_1 - T_2) \qquad (18\text{–}7)$$

Also, for the adiabatic reversible steps, AD and BC

$$
\left.\begin{array}{c}
\dfrac{p_A}{p_D} = \left(\dfrac{T_1}{T_2}\right)^{\frac{k}{k-1}} \\[20pt]
\dfrac{p_B}{p_C} = \left(\dfrac{T_1}{T_2}\right)^{\frac{k}{k-1}}
\end{array}\right\}
\quad \text{or} \quad \dfrac{p_A}{p_B} = \dfrac{p_D}{p_C}
\tag{18–8}
$$

Adding these work quantities for the whole four step cycle gives

equal and opposite, from Eq. 18–8

$$
\frac{|w_{\text{cycle}}|}{|q_1|} = \frac{|w_{AB}| + |w_{BC}| - |w_{CD}| - |w_{DA}|}{|q_1|}
\tag{18–9}
$$

$$
= \frac{R(T_1 - T_2)\ \ell n\ \dfrac{p_B}{p_A}}{RT_1\ \ell n\ \dfrac{p_B}{p_A}} = \frac{T_1 - T_2}{T_1}
$$

Miracle of miracles, this is exactly the same equation as obtained from Kelvin's work scale, Eq. 18–3.

This shows that the ideal gas temperature scale is equivalent to the Kelvin work scale, that $T = T'$, and that we can drop the prime. Because Kelvin made this discovery, we name our absolute temperature scale in his honor. So we can now proceed, confident that our ordinary ideal gas scale is consistent with the thermo-dynamic temperature scale.

Finally, returning to our usual notation of $-Q$ for heat lost from our Carnot engine, we can write for the engine

$$
\frac{-Q_2}{Q_1} = \frac{T_2}{T_1} \quad \text{or} \quad \frac{Q_2}{T_2} + \frac{Q_1}{T_1} = 0
$$

More generally, for any reversible Carnot engine that uses any number of sources and sinks at various temperatures, we can write

$$
\left.\begin{array}{c}
\dfrac{Q_1}{T_1} + \dfrac{Q_2}{T_2} + \dfrac{Q_3}{T_3} + \ \dots\ = 0 \\[20pt]
\sum \dfrac{Q_i}{T_i} = 0 \quad \text{or} \quad \displaystyle\int \dfrac{dQ_{\text{rev}}}{T} = 0
\end{array}\right\}
\tag{18–10}
$$

or

This quantity $\int dQ_{rev}/T$ always enters in cyclical reversible processes, it represents a change in a property of the system as is the enthalpy change or internal energy change, and we call it entropy change, ΔS.

So for any reversible change

$$\begin{array}{c}\textit{sources and sinks,}\\ \textit{system and surroundings,}\\ \textit{for everything}\\ \sum_{total}\Delta S = 0\end{array}$$

(18–11)

By introducing numbers we can show that for any irreversible change

$$\begin{array}{c}\textit{for everything,}\\ \textit{system and surroundings}\\ \sum_{total}\Delta S > 0\end{array}$$

(18–12)

C. THE IDEAL OR REVERSIBLE HEAT PUMP

By definition, a Carnot heat engine is a reversible engine, so if you run it backwards you are able to pump heat from a low temperature to a higher temperature. But this requires that the heat engine receives work, as shown in Figure 18–12.

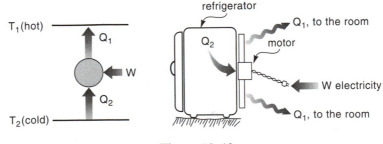

Figure 18–12.

For a refrigerator $|Q_1|$ is the heat entering the room

$|Q_2|$ is the heat removed from the interior of the refrigerator

$|W|$ is the electrical work needed to pump this heat "uphill".

In a home air conditioning-heating system the heat pump works one way in winter, the other way in summer, as sketched in Figure 18–13.

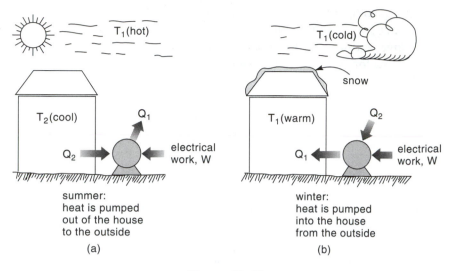

Figure 18–13.

The efficiency of operation of these units is measured in two different ways. For refrigerators or for heat pumps in summer we want to know how much heat can be removed from the cold region for each unit of work used. We call this ratio the *coefficient of performance*, or COP. Thus, from Eq. 18–3 and Figure 18–13a we have, ideally (reversible operations)

$$COP = \frac{|Q_2|}{|W|} = \frac{T_2}{T_1 - T_2} \qquad \text{\textit{for a refrigerator or}} \atop \text{\textit{a heat pump in summer}} \qquad (18\text{–}13)$$

For the heat pump used as a heater we are interested in the heat entering the house per unit of work used. Thus, from Eq. 18–3 and Figure 18–13b we have, ideally

$$COP' = \frac{|Q_1|}{|W|} = \frac{T_1}{T_1 - T_2} \qquad \text{\textit{. . . for a heater or a}} \atop \text{\textit{heat pump in winter}} \qquad (18\text{–}14)$$

Real heat pumps and refrigerators are less efficient and require more work to pump low temperature heat uphill, so

$$(COP)_{real} < (COP)_{ideal} \quad \text{and} \quad (COP')_{real} < (COP')_{ideal} \qquad (18\text{–}15)$$

D. THE T-S DIAGRAM FOR THE CARNOT CYCLE

The T-s diagram shows the Carnot cycle most clearly. Thus, per unit quantity of circulating gas, Figure 18–14 shows the Carnot heat engine for the work-producing device

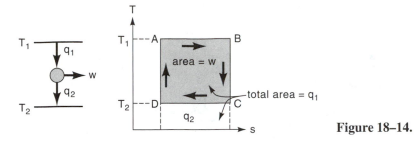

Figure 18–14.

For the Carnot refrigerator (work added to the circulating gas) Figure 18–15 shows

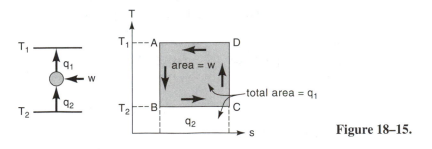

Figure 18–15.

These T-s diagrams represent Eq. 18–3.

E. FOR NONIDEAL HEAT ENGINES

What goes on in a Carnot engine? What makes it work? First, let's point out that it is an idealization, a conceptual engine that does not in fact exist. However, the circulating ideal gas system, discussed after Eq. 18–3, which follows a four-step cycle

- isothermal heat input at T_1
- adiabatic reversible expansion
- isothermal heat rejection at T_2
- adiabatic reversible compression

does represent a Carnot engine or heat pump.

In the real world the two adiabatic steps are not perfectly reversible, and the desired isothermal heat addition and heat removal steps are not really isothermal, so for the real heat engine we get a distorted rectangle on the T-s diagram, shown in Figure 18–16 in somewhat exaggerated form.

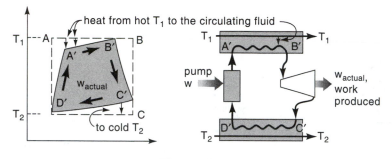

Figure 18–16.

This figure and its opposite for the heat pump illustrate the general finding on work produced (in the engine) or work needed (in the refrigerator).

for work produced (by engine)	$\|W_{irr}\| < \|W_{rev}\|$
for work needed (by heat pump)	$\|W_{irr}\| > \|W_{rev}\|$

(18–16)

or more generally

$$W_{rev} > W_{irr} \quad \text{and} \quad q_{rev} = T\Delta s > q_{actual}$$

(18–17)

Applied thermo shows how this cycle can be approximated, not just with a gas but with a fluid that evaporates, to absorb heat $|q_1|$ (to go from point A to point B on Figure 18–16), and that condenses to reject heat $|q_2|$ (point C to D) as it flows around the cycle. This fluid could be water, HFC-134a, NH_3, mercury, or one of the many other fluids, depending on the temperature range of interest.[2]

---

[2]Are you reading this paragraph under an electric light? If so, then chances are that the electricity needed to make this light work was generated in a giant power station operating with water boiling and steam condensing according to this very cycle.

EXAMPLE 18–1.   The Carnot heat engine

A large heat reservoir at 100°C is supplying heat to a large sink at 0°C. A Carnot engine operates between source and sink. Calculate the efficiency of this engine assuming that the temperature of source and sink remain unchanged.

Solution

Let us first sketch this system (Figure 18–17).

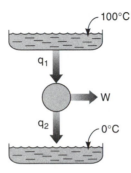

Figure 18–17.

From Eq. 18–3 the efficiency of the engine is

$$\eta = \frac{|W|}{|Q_1|} = \frac{T_1 - T_2}{T_1} = \frac{373 - 273}{373} = \underline{\underline{26.8\%}} \longleftarrow$$

EXAMPLE 18–2.   Another Carnot engine

100 kg of water at 100°C furnishes heat to a Carnot engine that discards heat to a sink consisting of 100 kg of cold water at 0°C. Source cools, sink heats, and eventually both end at the same temperature. Calculate

(a) the final temperature of the 200 kg of water

(b) the work obtainable

Solution

First, sketch the system (Figure 18–18).

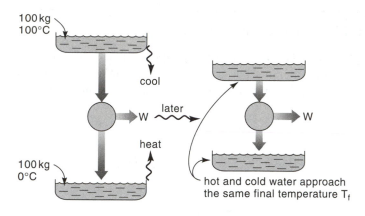

Figure 18–18.

(a) If we just cooled the hot water with the cold without doing work, then

$$\Delta H_{hot} + \Delta H_{cold} = 0$$

or $\qquad m_{hot}c_p(T_f - T_1) + m_{cold}c_p(T_f - T_2) = 0$

$$100(100 - T_f) + 100(0 - T_f) = 0$$

or $\qquad T_f = \dfrac{10000}{200} = 50°C$

But this is not the situation we are considering in this problem. We are doing work here, removing energy from the hot and cold water system, so the final temperature must be less than 50°C. But how do we find the final temperature? Here's the clue. Since we are doing this operation with a Carnot heat engine, we are doing it reversibly. Thus, with no heat interchange with the surroundings, just between hot and cold water, we have

$$\Delta S_{total} = \Delta S_{hot} + \Delta S_{cold} = 0$$

But $\qquad \Delta S_{hot} = \displaystyle\int_{T_1}^{T_f} \frac{mc_p dT}{T} = m_{hot}c_p \, \ell n \frac{T_f}{373}$

$$\Delta S_{cold} = \displaystyle\int_{T_1}^{T_f} \frac{mc_p dT}{T} = m_{cold}c_p \, \ell n \frac{T_f}{273}$$

Combining gives

$$m_{hot}c_p \, \ell n \frac{T_f}{373} + m_{cold}c_p \, \ell n \frac{T_f}{273} = 0$$

or $T_f = \sqrt{273 \cdot 373} = 319 \text{ K} = \underline{46°C}$ ⟵ a

(b) Work done is the energy lost in going from 50°C to 46°C. Thus

$$W = mc_p \, (50 - 46) = 200 \, (4184)(4) = 3.347 \times 10^6 \, J$$

$$= \underline{3.347 \text{ MJ}}$$ ⟵ b

EXAMPLE 18–3.   Still another Carnot engine (Figure 18–19)

100 kg of water at 100°C furnishes heat to a Carnot engine that discards heat to a large sink at 0°C. The process continues, the 100°C water cools and finally ends up at 0°C. Calculate the maximum amount of work obtainable.

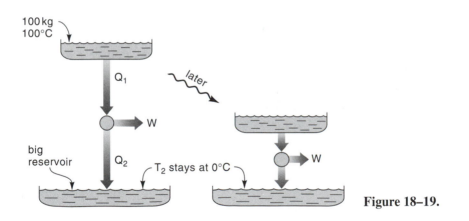

Figure 18–19.

Solution

Since this operation is done reversibly (with the Carnot engine), we write

$$\Delta S_{total} = \Delta S_{hot} + \Delta S_{cold} = 0 \qquad (18\text{–}18)$$

For the source that gives up heat

$$\Delta s_{hot} = \int \frac{dQ_1}{T_1} = \int_{T_1}^{T_2} \frac{mc_p dT_1}{T_1} = mc_p \, \ell n \frac{T_2}{T_1}$$

$$= 100(4184) \, \ell n \frac{273}{373} = -130 \, 585 \, J \qquad (18\text{–}19)$$

For the sink that gains heat but stays at 0°C

$$\Delta S_{\text{cold}} = \int\limits_{\text{at } T_2} \frac{dQ_2}{T_2} = \frac{Q_2}{T_2} = \frac{Q_1 - W}{T_2} = \frac{mc_p(T_1 - T_2) - W}{T_2}$$

$$= \frac{100(4184)(100 - 0) - W}{273} = 153\,260 - \frac{W}{273} \qquad (18\text{–}20)$$

Replacing Eqs. 18–19 and 18–20 in 18–18 gives

$$-130\,585 + 153\,260 - \frac{W}{273} = 0$$

or

$$W = 6190\,295 \text{ J} = \underline{6.19 \text{ MJ}} \quad \longleftarrow$$

EXAMPLE 18–4.   Heat pumps for vacation cabins

With an ideal heat pump determine how much energy can be pumped into a home, which is to be kept at 27°C, from a lake bottom whose water is at 7°C, for each kW·hr of electricity used (Figure 18–20). In effect calculate the coefficient of performance of this heat pump in this situation.

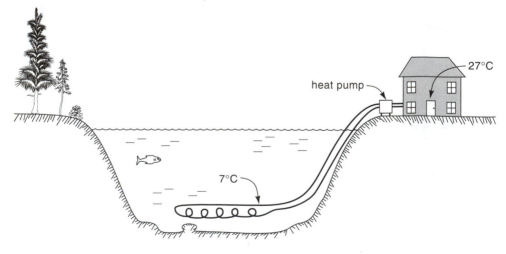

Figure 18–20.

Solution

Represent the ideal heat pump and its insulated piping with a Carnot engine (Figure 18–21).

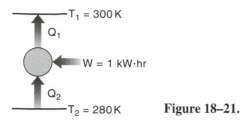

Figure 18–21.

We want to calculate $|Q_1|$ given $|W| = 1$. Write Eq. 18–3

$$\frac{|Q_1|}{|W|} = \frac{T_1}{T_1 - T_2}$$

Then

$$|Q_1| = \left(\frac{T_1}{T_1 - T_2}\right)|W| = \left(\frac{300}{300 - 280}\right)1 = 15 \text{ kW·hr}$$

Therefore, from Eq. 18–14, the coefficient of performance of this ideal device is

$$\underline{\underline{COP'}} = \frac{|Q_1|}{|W|} = \frac{15 \text{ kW·hr}}{1 \text{ kW·hr}} = \underline{\underline{15}} \quad \longleftarrow$$

Note. With an ideal heat pump you save 14/15, or about 93% on the cost of fuel. With actual commercial heat pumps, with their various inefficiencies

$$(COP')_{\text{actual}} \cong 3$$

PROBLEMS

1. Below I show two heat engines that are operating between the same two temperatures. I think that one of them is not a Carnot engine. If so, which one is it? Show your reasoning.

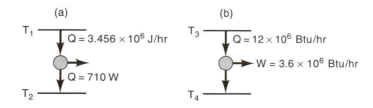

2. A Carnot heat engine continuously absorbs heat from a source at 227°C, puts out 400 W of power, and discards heat continually at 27°C. What is the rate of entropy change

   (a) for the source

   (b) for the sink

   (c) total, for the whole engine

3. The Carnot freezer in our 300 K laboratory has a COP of 11. How cold is it in the freezer?

4. A reversible heat engine receives heat from a source at 500 K and rejects heat to a sink at 300 K. What should be the heat input rate to power a 10 kW (of work) engine?

5. It has been proposed that Hell must be isothermal and absolutely flat (no hills). How can you defend such an argument?

6. A refrigeration machine requires 1 kW of power per ton of refrigeration.

   (a) What is its COP (coefficient of performance)?

   (b) How much heat is rejected to the condenser?

   (c) If the condenser operates at 15°C what is the lowest temperature the refrigerator can maintain assuming a heat load of 1 ton of refrigeration?

   Note. 1 ton of refrigeration = 12000 Btu/hr of heat abstracted from the cold reservoir.

7. A large heat reservoir at 100°C furnishes heat to a Carnot engine that discards heat to a sink consisting of 100 kg of water at 0°C. The process continues, the 0°C water heats and finally ends up at 100°C. Calculate the work obtained.

8. Here is some information on two Carnot engines. From this can you evaluate T_1 and T_3? If so, do so. If you feel that this is not possible, then say so.

9. A barge-mounted mini-OTEC (ocean thermal energy conversion) power station completed tests off the island of Hawaii (*Chemical and Engineering News*, 5 May 1980). The 50 kW turbogenerator ran on a closed ammonia cycle. This used the temperature difference between warm surface water (27°C), which cooled just a small fraction of a degree in the exchanger, and cold bottom water, which heated from 6°C to 7°C on the exchanger surface. If the operation was only 10% as effective as a Carnot cycle between these temperatures (that is, between 6–7°C and 27°C), find the needed pumping rate of cold water up from the ocean bottom.

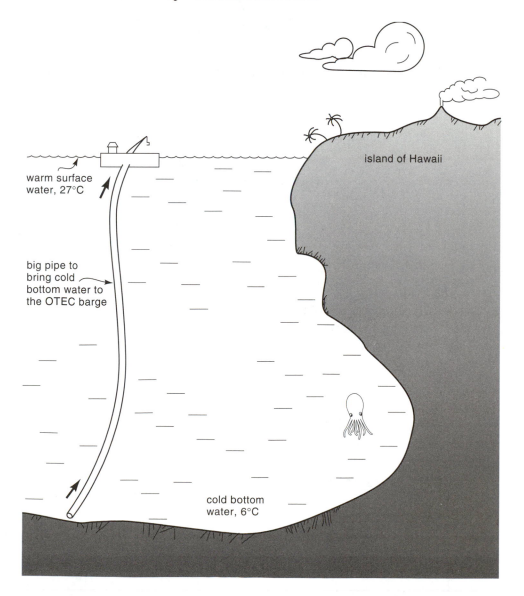

island of Hawaii

warm surface water, 27°C

big pipe to bring cold bottom water to the OTEC barge

cold bottom water, 6°C

10. A Carnot engine is connected to three reservoirs as sketched below. What is the power output of this engine?

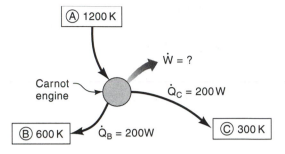

11. Given a Carnot engine and three reservoirs, as sketched. Find Q_1 and Q_3.

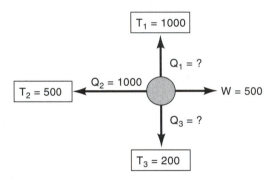

12. Most but not all home refrigerators are electrically powered. The Servel-Electrolux is one of the few exceptions since it is powered by natural gas, in fact, by the heat of a gas flame. In terms of Carnot cycles it operates ideally as follows:

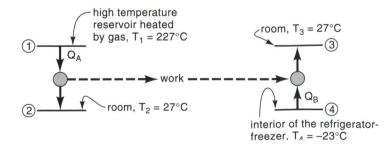

How much heat must the high temperature reservoir give up, Q_A, to extract 1 unit of heat, Q_B, from the refrigerator?

13. Suppose high pressure 300 K air enters an ideal Hilsch tube and leaves in two streams, one at 200 K and 1 bar, the other at 400 K and 1 bar. Sketch and explain (without numerical calculations) how you could reverse the process; that means take air at 200 K and 400 K at 1 bar and end up with a high pressure 300 K stream of air.

14. Close to ZigZag, OR, by the side of a dormant volcano, is a large 1 km³ underground field of hot fractured quartz-like rock (ρ = 2650 kg/m³, c_p = 800 J/kg·K). The average temperature of the rock is 500K, of surroundings is 280K. I wonder how much useful energy is stored in that rock.

(a) To estimate this, calculate how long I could produce 1000 MW of electricity on extracting heat energy from the rock if I did this most efficiently. Note, this power production rate is equivalent to the output of a large coal-fired power station.

(b) If we can sell electricity to the power company at 3¢/kW·hr, how much is this energy source worth?

CHAPTER 19

EXERGY OR AVAILABILITY

By now you may realize that much or most of thermo deals with work—all sorts of work: work obtainable from or work that has to be done for a given change. Let us briefly review these.

- First, from mechanics we saw that work was defined as (force) × (distance).
- Then with the first law we showed that the work done by a batch or closed system could be of two types: useful or what we call shaft work, W_{sh}, and pV work, which is what has to be done when the system itself expands and pushes back the atmosphere.
- In steady state flow processes the system doesn't expand or contract so there is no pV work associated with the system; however, the flow streams can have very different densities (liquid water may enter, steam may leave), so there is pV work associated with the flow streams pushing back the atmosphere. We combine this work with the internal energy, U, to give the enthalpy $H = U + pV$.
- When dealing with pumps and compressors, we talk of adiabatic reversible work, W_{rev}. We may think of this as the most efficient way of operating these devices, but this is not necessarily so. For illustration, Example 16–2 shows that an isothermal turbine can produce more work than the adiabatic reversible (114 kW versus 84 kW).
- We also saw that the work produced in reversible operations is usually, but not always, greater than in irreversible operations.
- And, finally, from the second law we saw how much work could be produced, not by high pressure fluid going to low pressure, but by heat flow from high to low temperature.

One would think that we've had enough of work in its various forms, but no. In this chapter we consider one more situation—the maximum work obtainable for a given change of the system.

Let us make clear what we are trying to explain. For this consider the maximum amount of work obtainable when a system changes from state 1 to state 2 in surroundings that are at state 0. For example, suppose a system changes from 420°C to 380°C during removal of heat and while doing work, in a room at 25°C, as shown in Figure 19–1.

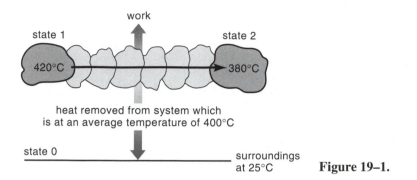

Figure 19–1.

But this is not the most work that we can extract, because the heat is removed at high temperature. We could get extra work as this heat flows "downhill" to 25°C by inserting a Carnot engine in the heat flow stream, shown in Figure 19–2.

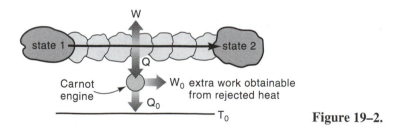

Figure 19–2.

This example shows that to get the true maximum available work, all the heat to be rejected from the system has to be rejected at the temperature of the surroundings.

In a similar manner, if the system and surroundings are at different pressures and if the system expands, then to minimize the nonuseful pV work, the expansion should always be done at the pressure of the surroundings. Any different pressure should be reduced to p_0 by producing useful work.

That, then, is the business of this chapter: to calculate the maximum extractable work, or the very least that needs to be done for a given change. This work depends on three quantities: the states of the system, 1 and 2, plus the state of the surroundings, 0. This work concept was first mentioned by J. C. Maxwell in his *Theory of Heat* (Longmans Green, London, 1871). It has been called many things:

- Maximum extractable work   ⎫
- Maximum available energy   ⎬ *We have used these terms*
- Maximum obtainable work   ⎭ *in the discussion above*
- Available work
- Available energy
- Availability

To distinguish it from all other forms of work, Rant (*Forsch. Ing-Wes.* 22, 36 [1956]) coined a new term, "exergy," to represent this concept.

In the United States the term "availability," with symbol B and b, is favored and used. It was first introduced by Gibbs, but developed and amplified by Keenan (*Thermodynamics*, Wiley, New York, 1941). In Europe and Japan the term "exergy" is favored.

Call it what you will, in this book we will call it "exergy." Remember, exergy represents not just any form of energy, but the maximum shaft work available in a system that is above some ground state.[1] We give it the symbol W_{ex}.

A.   EXERGY OF BATCH SYSTEMS, $W_{ex,batch}$

Let us develop the equations for exergy for various situations.

$W_{ex,1 \to 0,batch}$

Consider a system at T_1, p_1 and having E_{p1} and E_{k1}, while the surroundings are at T_0, p_0. To calculate the exergy of this system extract all the work that you can from it as you move it to T_0, p_0, where $E_{p0} = E_{k0} = 0$. This process is shown and explained in Figure 19–3. Study this figure. If you understand it, you've got the idea of exergy.

---

[1]Wouldn't it be mind boggling if we could evaluate the maxima possible in other areas? For example,
- The fastest possible run of the mile by man is 3 min 16 s.
- The highest that man could jump is 3.07 m.
- The longest possible life span of man is 186 years 42 days.

It is inconceivable that such figures can be calculated; however, in thermo we can evaluate such maxima. They are very useful for they allow us to compare actual processes with these ultimates, evaluate efficiencies, and see whether it is worthwhile putting research effort in trying to improve actual processes.

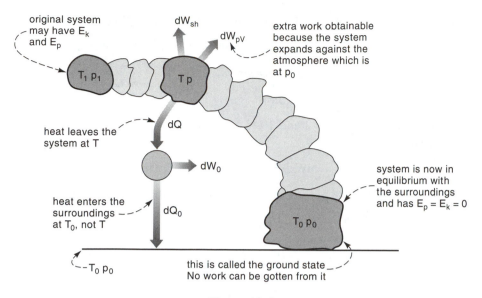

Figure 19–3.

Now let us develop the equation for this change from T_1, p_1 to T_0, p_0. First, from the Carnot engine analysis of the previous chapter we can write

$$\underset{\underset{\text{given up by the system}}{\big\downarrow}}{\overset{\overset{\text{received by surroundings}}{\big\downarrow}}{\frac{d|Q_0|}{d|Q|}}} = \frac{T_0}{T}$$

or

$$d|Q_0| = T_0 \,\frac{d|Q|}{T} = T_0 dS \qquad (19\text{–}1)$$

where the $\frac{d|Q|}{T}$ is *of system* and $T_0 dS$ is *of surroundings*.

Next, consider a very small (or differential) move of the system (at T,p) towards equilibrium. The total work produced is the sum of three terms (Figure 19–3).

$$dW_{total} = dW_{sh} + p_0 dV + dW_0 \qquad (19\text{–}2)$$

work in pushing back the atmosphere, dW_{pv}

we call the sum of these two terms the available shaft work, useful work, or the exergy, $dW_{sh,av}$

work obtained from the Carnot engine (Figure 19–3)

With the first law, Eq. 3–2, we have for the system plus Carnot engine

$$d\mathbf{E} = dQ_0 - dW_{total}$$

*total change in
energy of the
system* $\;\; T_0\, dS \ldots from\; Eq.\; 19–1$ (19–3)

Combining Eqs. 19–2 and 19–3 gives the maximum available shaft work, or exergy, as

$$dW_{ex} = dW_{sh} + dW_0 = -d\mathbf{E} + T_0 dS - p_0 dV \tag{19–4}$$

So for the whole progression of changes for the system from T_1, p_1, E_{p1}, E_{k1} to T_0, p_0, with $E_{p0} = E_{k0} = 0$, Eq. 19–4 becomes

$$\boxed{W_{ex,1\to0,\text{batch}} = -(U_0 - \mathbf{E}_1) + T_0(S_0 - S_1) - p_0(V_0 - V_1) \quad [J]}$$

$$U_1 + E_{p1} + E_{k1}$$

(19–5)

*This is the maximum useful shaft work that can be squeezed out of the system,
including electrical, magnetic, kinetic, chemical, potential, and so on, as it
moves to equilibrium or the dead state. It is the exergy of the system at state 1.*

$W_{ex,1\to2,\text{batch}}$

When a batch of material goes from state 1 to state 2, with surroundings at state 0, we have the situation sketched in Figure 19–4.

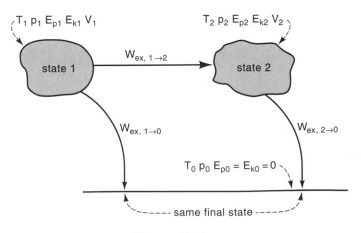

Figure 19– 4.

From Figure 19–4 we see that

$$W_{ex,1\to2} = W_{ex,1\to0} - W_{ex,2\to0} \tag{19-6}$$

Hence, combining Eq. 19–4 with Eq. 19–6 gives

$$\boxed{W_{ex,1\to2,\text{batch}} = -(\underbrace{E_2 - E_1}) + T_0(S_2 - S_1) - p_0(V_2 - V_1) \quad [\text{J}]}$$
$$\phantom{W_{ex,1\to2,\text{batch}}}\quad U_2 + E_{p,2} + E_{k,2} \tag{19-7}$$

Actual and Lost Work in Real Changes, Batch System

The actual work, including pV work received in the surroundings when a batch goes from state 1 to state 2, is given by

both useful and not useful

$$\Delta E_{1\to2} = Q_{\text{actual}} - W_{\text{actual}}$$
$$\qquad\qquad\text{to surr}\qquad\text{to surr} \tag{19-8}$$

$T_0\,\Delta S_{\text{surroundings}},$ *from Eq. 19–1* $\qquad W_{sh,actual,1\to2} + p_0(V_2 - V_1),$ *from Eq. 19–2*

Then the lost shaft work, from Eqs. 19–7 and 19–8 is

$$W_{sh,lost,1\to2} = W_{ex,1\to2} - W_{sh,actual,1\to2}$$
$$= T_0(S_2 - S_1) + T_0\Delta S_{surr}$$
$$= T_0\Delta S_{syst} + T_0\Delta S_{surr}$$

So for any change

$$\boxed{W_{sh,lost} = T_0\Delta S_{total} \quad [\text{J}]}$$
$$\phantom{W_{sh,lost}}\quad \textit{system + surroundings} \tag{19-9}$$

Equations 19–5, 19–7, and 19–9 are the three important exergy equations for batch systems.

EXAMPLE 19–1.   Prosperity comes to East Zilch (Figure 19–5)

The little community of East Zilch, TX, (pop ~ 35, average temperature $\cong 27°C$) owns an enormous underground reservoir—not of oil, not of natural gas, but of waste gas (volume of gas in reservoir = 10^{12} m³, c_p = 36 J/mol·K, p = 9.95 atm, T = 237°C, $\overline{mw}$ = 0.03 kg/mol). The residents feel that there surely should be some extractable energy (and some money) in this gas, so they hire you as a consultant to answer the following questions.

   a. Ideally, how much useful work can be gotten from this gas?
   b. If East Zilch can sell electricity to the power company at 3.6¢/kW·hr, and if their energy extraction plant is 10% efficient, what is the value of the gas in the reservoir?

Solution

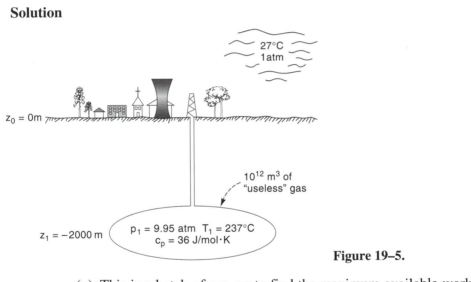

Figure 19–5.

   (a) This is a batch of gas, so to find the maximum available work when this gas expands from $p_1 T_1$ to $p_0 T_0$, use Eq. 19–5

$$W_{ex,1\to0} = -\left[(U_0 + \cancel{E_{p,0}} + \cancel{E_{k,0}}) - (U_1 + E_{p,1} + \cancel{E_{k,1}})\right] + T_0(S_0 - S_1) - p_0(V_0 - V_1)$$

(19–10)

the reservoir is underground, ΔE_p is certainly different from zero, but assume it to be small and negligible

Assuming ideal gas behavior, the expression becomes

$$W_{ex,1\to 0} = -\left[n\, c_v (T_0 - T_1)\right] + T_0 n\left(c_p\, \ell n \frac{T_0}{T_1} - R\, \ell n \frac{p_0}{p_1}\right) - p_0(V_0 - V_1)$$

↖ *this is the exergy*   *Eq. 16–5* ⤵

(19–11)

where

$$n = \frac{p_1 V_1}{R T_1} = \frac{9.95\,(101\,325)(10^{12})}{(8.314)\,(510)} = 2.378 \times 10^{14} \text{ mol}$$

$$c_v = 36 - 8.314 = 27.686 \text{ J/mol·K}$$

$$V_0 = (10^{12})\left(\frac{9.95}{1}\right)\left(\frac{300}{510}\right) = 5.853 \times 10^{12} \text{ m}^3$$

Replacing all known values into Eq. 19–11 gives

$$W_{ex,1\to 0} = -2.378 \times 10^{14}(27.686)(300 - 510) + 300\left(2.378 \times 10^{14}\right)$$

$$\left[36\, \ell n \frac{300}{510} - 8.314\, \ell n \frac{1}{9.95}\right] - 101\,325(5.853 \times 10^{12} - 10^{12})$$

$$= 1.3826 \times 10^{18} - 4.2722 \times 10^{13} - 4.9173 \times 10^{17}$$

$$= \underline{\underline{8.9083 \times 10^{17} \text{ J}}} \text{ useful work in this gas} \quad \longleftarrow$$

(b) Value of this gas if 10% of all this available work could be converted to electricity which is then sold

$$(8.9083 \times 10^{17} \text{ J})\left(\frac{1 \text{ kW}}{1000 \text{ J/s}}\right)\left(\frac{1 \text{ hr}}{3600 \text{ s}}\right)\left(\frac{\$0.036}{\text{kW·hr}}\right)(0.1) = \$8.9 \times 10^8$$

$$= \underline{\underline{\$890 \text{ million}}} \quad \text{WOW!!} \quad \longleftarrow$$

EXAMPLE 19–2. More on East Zilch

Repeat Example 19–1 but do not ignore the potential energy contribution (the work required) to bring the gas to the earth's surface.

Solution

The potential energy contribution, ignored in Example 19–1, is

$$E_{p,1\to0} = E_{p0} - E_{p1}$$

$$= \frac{mg(z_0 - z_1)}{g_c}$$

$$= \frac{(2.378 \times 10^{14})(0.03)(9.8)(0 - (-2000))}{(1)}$$

$$= 1.3983 \times 10^{17} \text{ J}$$

So, accounting for the work needed to bring this gas to the surface lowers W_{ex} somewhat. It is then

$$W_{ex,1\to0} = 8.9083 \times 10^{17} - 1.3983 \times 10^{17} = 7.51 \times 10^{17} \text{ J}$$

Thus, the available work is lowered by

$$\frac{1.3983}{8.9083} \times 100 = 15.7\%$$

and the value of this gas is lowered by

$$\$890 \times 10^6 \ (0.157) = \underline{\$140 \text{ million}} \quad \longleftarrow$$

This example shows that the potential energy term should not always be ignored.

B.   EXERGY IN FLOW SYSTEMS

Consider the flow system shown in Figure 19–6.

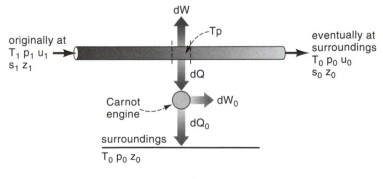

Figure 19–6.

Here we have the same argument as with the batch system, that extra work can be extracted by use of a Carnot engine if whatever heat to be rejected reaches the surroundings at T_0. However, pushing back the atmosphere does not have to be accounted for by an extra term, as in batch systems. It is already accounted for by use of the enthalpy, instead of internal energy (see above Eq. 13–3).

Thus, the exergy, or shaft work available, of the flowing stream in state 1 (see Eq. 19–5) is

$$W_{ex,1\rightarrow 0} = -\left[H_0 - (H + E_p + E_k)_1\right] + T_0(S_0 - S_1) \tag{19–12}$$

The exergy as the flowing system goes from state 1 to state 2 (similar to Eq. 19–7) is

$$W_{ex,1\rightarrow 2} = -\left[(H + E_p + E_k)_2 - (H + E_p + E_k)_1\right] + T_0(S_2 - S_1) \tag{19–13}$$

and the lost or wasted work for an actual change between state 1 and state 2 (see Eq. 19–9) is

$$W_{sh,lost,1\rightarrow 2} = W_{ex,1\rightarrow 2} - W_{sh,actual,1\rightarrow 2}$$

$$\left[Q_{rev} - (\Delta H + \Delta E_p + \Delta E_k)\right] \nearrow \qquad \left[Q_{\substack{actual\ to \\ surroundings}} - (\Delta H + \Delta E_p + \Delta E_k)\right]$$

$$= Q_{rev} - Q_{actual} \tag{19–14}$$

$$= T_0(S_2 - S_1) + T_0 \Delta S_{surr}$$

So for an actual change from state 1 to state 2

$$W_{sh,lost,1\rightarrow 2} = T_0 \Delta S_{total} \tag{19–15}$$

for flow streams and for surroundings

C. RELATIONSHIP BETWEEN WORK TERMS IN BATCH OR FLOW SYSTEMS

When a system changes from state 1 to state 2 we have, in general,

Batch . . . . Eq. 19–5 . . . . Eq. 19–8 . . . Eq. 19–9

$$W_{ex,1\rightarrow 0} = W_{ex,2\rightarrow 0} + W_{\substack{sh,actually \\ done, \\ 1\rightarrow 2}} + W_{sh,lost,1\rightarrow 2} \tag{19–16}$$

Flow system . . . . Eq. 19–12 . . . Eq. 19–14 . . . Eq. 19–15

EXAMPLE 19–3.   Exergy of an ideal gas (Figure 19–7)

A stream of 2 mol/s of air goes from 1000 K and 10 bar to 500 K and 5 bar while doing 5.0 kW of work. Surroundings are 300 K and 1 bar. What is the lost work for this process?

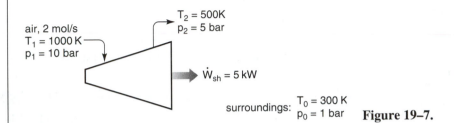

air, 2 mol/s
T_1 = 1000 K
p_1 = 10 bar

T_2 = 500K
p_2 = 5 bar

$\dot{W}_{sh}$ = 5 kW

surroundings: T_0 = 300 K
p_0 = 1 bar   **Figure 19–7.**

Solution

Let us calculate the exergy of the incoming minus outgoing flow streams and then subtract the actual work done. This should give the lost work. So from Eq. 19–13 we have per mol of flowing gas

$$w_{ex,1\rightarrow2} = h_1 - h_2 + T_0(s_2 - s_1) \qquad \text{Eq. 16–5}$$

$$= c_p(T_1 - T_2) + T_0 \left[c_p \, \ell n \frac{T_2}{T_1} - R \, \ell n \frac{p_2}{p_1} \right]$$

$$= 29.1 \, (1000 - 500) + 300 \left[29.1 \, \ell n \frac{500}{1000} - 8.314 \, \ell n \frac{5}{10} \right]$$

$$= 10\ 228 \text{ J/mol}$$

The available power is then

$$\dot{W}_{ex,1\rightarrow2} = (10\ 228)\ (2) = 20\ 456 \text{ W}$$

$$= 20.5 \text{ kW}$$

The actual power production is $\dot{W}_{actual}$ = 5 kW, so

$$\dot{W}_{sh,lost} = 20.5 - 5.000 = \underline{\underline{15.5 \text{ kW}}} \quad \longleftarrow$$

EXAMPLE 19–4.   Showers for Saudi Arabians

Saudi Arabia, a hot dry country ($T_{ave} = 30°C$), plans to lasso icebergs in Antarctica, tow them to Jiddah harbor, melt and store the water at 5°C, and thereby supply the country with fresh water (Figure 19–8). But one can produce work, electricity, and air conditioning in addition to fresh water during the melting process. If they do not try to recover this available work, how much power do they waste if they bring in a 10^6 ton iceberg every three weeks?

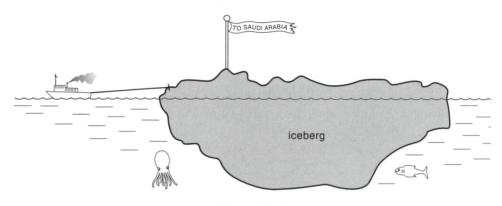

Figure 19–8.

Solution

The melting of ice followed by heating of water can be represented as Figure 19–9.

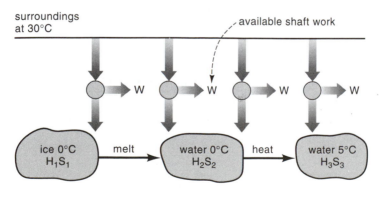

Figure 19–9.

Since we are only interested in the overall 1→3 change, we can bypass state 2. Also, consider this to be a steady state operation, and ignore ΔE_p and ΔE_k, in which case Eq. 19–13 for each kg of ice becomes

$$w_{ex,1\to3} = -\left[(h_3 + e_{p3} + e_{k3}) - (h_1 + e_{p1} + e_{k1})\right] + T_0(s_3 - s_1) \qquad (19–17)$$

Here we could use latent heats and heat capacities to calculate the values of enthalpies and entropies. More simply, just look up the tables at the back of this book. This gives us

$$h_1 = -333.43 \text{ kJ/kg} \qquad h_3 = 20.98 \text{ kJ/kg}$$
$$s_1 = -1.221 \text{ kJ/kg·K} \qquad s_3 = 0.0761 \text{ kJ/kg·K}$$

Hence, Eq. 19–17 becomes

$$w_{ex,1\to3} = -\left[(20.98 - (-333.43))\right] + 303(0.0761 - (-1.221))$$

$$= 38.61 \text{ kJ/kg}$$

Now the lost work is given by Eq. 12–14 as

$$w_{lost} = w_{ex} - \overbrace{w_{actually \atop extracted}}^{= 0} = 38.61 \text{ kJ/kg}$$

So the power lost is

$$\dot{W}_{lost} = \left(38.61 \ \frac{kJ}{kg}\right)\left(\frac{1000 \text{ kg}}{\text{ton}}\right)\left(\frac{10^6 \text{ tons}}{21 \text{ days}}\right)\left(\frac{1 \text{ day}}{24 \times 60 \times 60 \text{ s}}\right)$$

$$= 21\,280 \ \frac{kJ}{s} = \underline{\underline{21.3 \text{ MW}}} \quad \longleftarrow$$

This represents about 14 000 portable electric heaters or 20 000 electric coffee pots, all working continuously, day and night.

Note. See *Iceberg Utilization*, A. A. Husseiny, ed. (Pergamon, 1978) for discussions on how to capture and tow icebergs, best time of year for capture, best travel route, and thermo suggestions on how to coax icebergs to propel themselves, without towlines and tugs!

Chemical Engineering News, August 6, 1973, pg. 40, also considers the problems of supplying San Diego with iceberg water.

This chapter is the capstone chapter on work. It brings together various kinds of work, W_{pv}, W_{sh}, W_{rev}, W_{lost}, W_{ex}, so we have a large collection of problems to follow.

Note that throughout this chapter we have replaced Q by TΔS. This is to indicate that we are talking of reversible processes—with no friction and no wasted work. We are always trying to squeeze as much work from the system as possible.

For more on exergy analysis with bountiful examples from physical, chemical, and combustion operations see the minibook *Availability (Exergy) Analysis* by M. V. Sussman (Lexington, MA: Mulliken House, 1980).

PROBLEMS

1. Block A of iron (200 kg) is 10 m above ground level, block B (100 kg) is 20 m above ground level. Which block has a higher exergy?

2. Two blocks of iron (c_p = 500 J/kg·K) sit on the ground. Block A (200 kg) is at 400 K, block B (100 kg) is at 500 K. The ground and surroundings are at 300 K. Which block has the higher exergy?

3. Which has the higher exergy, block A of iron (100 kg, c_p = 500 J/kg·K) 10 m up in the air and at ambient temperature, or block B (100 kg) resting on the ground but at 10°C above the temperature of the surroundings, which are at 300 K?

4. A 2 liter plastic pop bottle is filled with air to 12.5 bar. The temperature is 300 K everywhere. What is its exergy? Does your calculated value make sense when compared with the values found in Examples 11–6, 11–7, and 11–8?

5. A big chemical plant is venting 3.6 tons/hr of clean low pressure steam (200°C and 1 bar) into the surroundings (25°C, 1 bar). What a waste!! What is the maximum recoverable power from this stream?

6. During full production of plutonium, the Hanford atomic energy facility in Washington continuously took in and discharged about 10^5 m³/hr of Columbia River water for cooling purposes. The river and surroundings are at an average temperature of 10°C, the discharged water was at 90°C. Theoretically, how much power could have been recovered from this hot water stream?

7. 2 kg of hot water (90°C) are poured into a swimming pool (20°C). Calculate

   (a) the total entropy change for this process, ΔS_{total}

   (b) the lost work, W_{lost}

   (c) the exergy of these 2 kg of hot water

8. Geothermal steam at 500 kPa and 250°C can be piped out of the ground at Klamath Falls, OR, at 1000 kg/min. With surroundings at 25°C how much power can be generated from this source?

9. An ice plant is to produce 100 metric tons of ice per hour in a continuous process. How many Dodge straight six engines rated at 225 HP each are needed to do the job if water is taken in at 20°C, cooled to 0°C and then frozen?

   Data   Heat is rejected at 20°C.

   Overall efficiency of the plant is 20%.

10. Imagine feeding a well insulated Hilsch tube with high pressure gas (3.3294 bar, 300 K, c_p = 34.026 J/mol·K) in a laboratory at 1 bar. Suppose the flow split is 50-50% with cold gas leaving at 100 K and hot gas leaving at 500 K. Is it possible to get this in theory or not?

11. What is the theoretical minimum amount of work needed to separate a stream of air into two streams, one of pure oxygen (21%), the other of nitrogen plus trace gases (79%), all this to take place at 300 K and 1 bar?

12. A stream of air (100 mol/s, 78% N_2, 21% O_2, 1% Ar) at 227°C and 1 bar is to be separated into two streams, one of pure argon, the other with no argon at all, both streams at 227°C and 1 bar. What is the minimum power needed for this operation?

13. A 25 kg block of copper (c_p = 400 J/kg·K) is cooled from 177°C to 27°C in Northern Kamchatka where the air temperature in summer is −23°C.

    (a) What is the largest amount of useful work that could have been recovered in this process?

    (b) Determine the amount of heat entering the surroundings when no work is done and when the useful work of part (a) is done.

14. A heat reservoir consisting of 100 kg H_2O at 100°C furnishes heat to a Carnot engine that discards heat to a large sink at 0°C. The process continues until the temperature of the original water cools to 0°C. Calculate the work obtainable.

15. A large heat reservoir that stays at 100°C supplies heat to a Carnot engine that discards heat to a sink consisting of 100 kg of water at 0°C at the start. The sink then rises in temperature eventually reaching 100°C. Calculate the work obtainable. Solve two different ways

(a) With a Carnot engine analysis of Chapter 18

(b) With an exergy analysis of this chapter

> *Note:* When only heat is involved in the process (no compressors or pressure changes), then one can use either the analysis of Chapter 18 or the methods of this chapter to solve problems.

16. An ideal gas (c_p = 36 J/mol·K) at 500 K and 1 atm is compressed to 10 atm, and then after suitable heat interchange it is piped to a process. This whole operation takes place in surroundings at 300 K. The actual pump work needed for this operation is about 20 kJ/mol of gas. Compare this value with

(a) The minimum work required for an adiabatic reversible compression of gas. Note that the final temperature of the gas will be higher than 500 K.

(b) The minimum work required for an isothermal compression. Here the final temperature will be 500 K.

(c) The very minimum work needed to get the gas to 10 atm and 500 K.

(d) If the product gas is to be delivered at 10 atm and 300 K, how does this affect your answer to part (c)?

17. 200 mol/s of combustion gas (k = 1.3) leaves a pressurized coal combustor at 1000 K and 10 bar. This waste gas goes through an adiabatic reversible turbine, exhausts at 2 bar and then is discarded to the atmosphere (300 K, 1 bar).

(a) At what temperature does the gas leave the turbine? Also calculate c_p of this gas.

(b) What is the power output of the turbine?

(c) What is the power lost in this operation compared to the best work extracting system?

18. Here you are, venting 10 lit/s of 400 K, 10 bar air. What a waste! Why not try to recover something from it? How much power can be recovered

(a) with an adiabatic reversible turbine, and

(b) with a 100% efficient isothermal turbine?

(c) What is the maximum available extractable power?

19. A stream of ideal gas (k = 1.333) at 16 bar and 720 K is expanded down to 1 bar and is cooled to 360 K without any useful work being produced.

(a) What is the c_p of the gas?

(b) What is the heat interchange per mole of gas with surroundings during this operation?

(c) What is ΔS of the gas for this operation?

(d) What is the maximum available work extractable from the incoming gas, the exergy, if surroundings are at 1 bar and 300 K?

20. *Chemical and Engineering News*, 21 April 1980, pg. 34, tells of power production from geothermal brine coming from the East Mesa geothermal field in California's Imperial Valley ($T_{ave} = 27°C$). In the U.S. Department of Energy's pilot plant 750 lit/min of hot brine at 170°C and under pressure enters the top of a 12 m high tower, 1 m in diameter, transfers its heat to isobutane, the working fluid, and then leaves at 65°C. At the same time liquid droplets of isobutane enter the tower, absorb heat, vaporize, leave the tower to feed a turbine to generate useful work, are then condensed, compressed (negligible energy required because $\Delta v \cong 0$) and returned to the tower. *Chemical and Engineering News* reports that the power output of the pilot plant is 500 kW. What is the efficiency compared to the Carnot efficiency? In the absence of thermodynamic data for brine take its properties to be that of water.

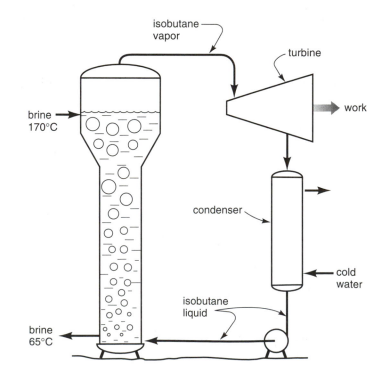

21. *Corvallis Gazette Times*, 27 May 1980, pg. 8, reports that tests on geothermal wells drilled near the Los Alamos Scientific Laboratories in New Mexico show that energy derived from hot rocks deep in the earth could be quite economical. It is done as follows.

- Drill holes 3000 m or more through impermeable rock down to a zone of hot fractured rock.
- Pump room temperature water (25°C) down at pressure and retrieve it as hot water.
- Run the hot water through heat exchangers or power turbines to extract useful work.
- Return the wastewater back underground to complete the circulation system.

If the water for a typical well is circulated at 10 kg/s and comes out of the ground as liquid at 200°C, what is the maximum power output of the well? Ignore the pumping power needed to pump water up, down, around the loop.

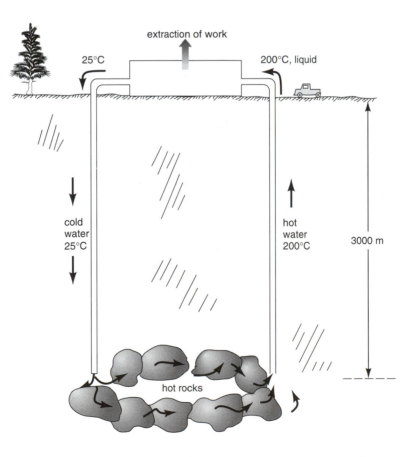

20

THERMO IN MECHANICAL ENGINEERING

The widest users of thermo are the mechanical engineers, for they are the designers and builders of engines, devices for transforming heat and chemical energy into work. They also deal with engines that use work to pump heat from a lower to a higher temperature. Let us look at these devices.

A. TYPES OF ENGINES

There are three broad types of engines for transforming heat into work; closed-cycle G-L engine, one-pass open-cycle G-L engine, and one-pass open-cycle gas engine.

Closed-Cycle G-L Engine (Figure 20–1)

This class of engine circulates a fluid, heating and vaporizing it at high pressure, expanding it, then cooling and condensing it, all in such a way as to generate work.

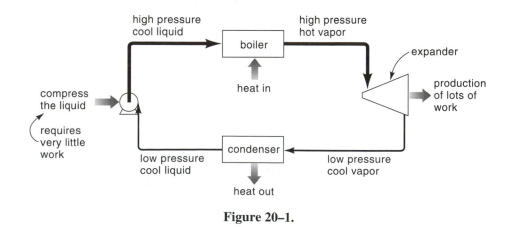

Figure 20–1.

These devices need both boiler and cooler-condenser. Large stationary coal-fired power plants for producing electricity are of this type, using liquid water and steam as the working fluid.

One-Pass Open-Cycle G-L Engine (Figure 20–2)

This type of engine compresses, heats and vaporizes a liquid, expands it to do work, and then discards the final low pressure vapor.

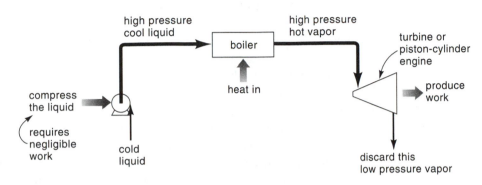

Figure 20–2.

Since the working fluid is only used once and then discarded, this type of engine is only practical when it can use a cheap fluid—that means water/steam. It is impractical to use any other fluid such as Freon or HFC-134a. This engine does not have a cooler-condenser so can be much smaller than the type 1 engine. The steam locomotive is the most romantic example of this type of engine.

One-Pass Open-Cycle Gas Engine (Figure 20–3)

The gas engine dispenses with both boiler and cooler-condenser. It generates the hot high pressure gas by a chemical reaction—combustion.

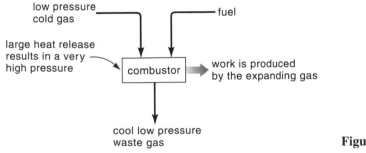

Figure 20–3.

As we will see there are many ways of doing this combustion—in piston-cylinders, in turbines, and others.

This type of engine can be very small and compact but requires using a liquid combustible fuel to generate the heat—gasoline, diesel, jet fuel. The automobile engine, the jet engine, the portable leaf blower, and the rocket engine are examples of this type of work producer. As long as liquid fuels are plentiful and cheap, this class of engine will dominate the world of engines.

Refrigerators and Heat Pumps

For pumping heat from a low temperature to room temperature one cannot use water/steam, one must use a fluid that boils and condenses at lower temperature. This leads us to ammonia, sulfur dioxide, Freons, and recently to hydrofluorocarbons, the HFCs. One-pass operations are not practical with these fluids, so we must use an engine that does not discard the working fluid, but uses a boiler and a condenser to recirculate and reuse the working fluid again and again.

In the past these engines were giant monsters serving commercial refrigeration plants. But with modern technology, smaller units still retaining boiler and condenser became practical; for example, the home refrigerator in the 1930s, the window air conditioner in the 1950s, and the really compact automobile air conditioner in the 1970s. Today we accept these devices as a matter of course, without any appreciation of their complexity. For very small scale operations we have other clever devices to "produce cold."

Let us now look at the above mentioned engine types in more detail and compare their operations with the ideal Carnot engine.

B.   THE CARNOT CYCLE

Chapter 18 shows that the Carnot cycle is the most efficient way to produce work from the flow of heat from high temperature T_1 to low temperature T_2. In general, the efficiency of work production is defined as

$$\eta = \frac{\text{work performed}}{\substack{\text{heat from} \\ \text{high temperature source}}} = \frac{|W|}{|Q_1|} = \frac{|Q_1| - |Q_2|}{|Q_1|} \qquad (20\text{–}1)$$

For the Carnot cycle this efficiency is dependent only on the high temperature of the source and low temperature of the sink, or

$$\eta = \frac{T_1 - T_2}{T_1} \qquad (20\text{–}2)$$

This cycle is represented by a rectangle on the T-s diagram, as shown in Figure 20–4 (also see Figures 18–11, 18–14, and 18–15).

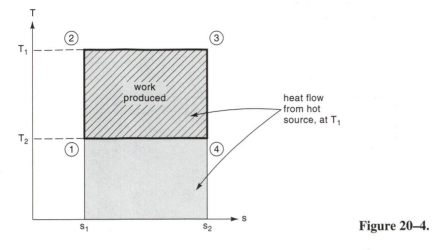

Figure 20–4.

Now the Carnot cycle can be well approximated by the circulation of a gas (see Chapter 18), but not by gas-liquid systems. However, the most useful and easily understood way of comparing practical engine cycles with the Carnot cycle is with the T-s diagram. So let us keep this figure in mind as we develop the various practical two-phase cycles.

C. PRACTICAL G-L CYCLES: THE RANKINE (POWER PLANT) CYCLE

Let us try to operate a Carnot cycle with a real fluid, say water/steam. To absorb or reject heat at constant temperature (steps 2→3 and 4→1 of Figure 20–4) can only be done by operating in the two-phase region, as shown in Figure 20–5.

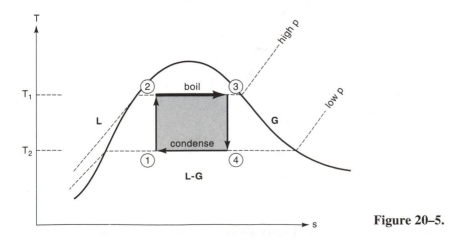

Figure 20–5.

But mechanical problems such as cavitation and erosion of turbine blades make it impractical to operate compressors and turbines in the two-phase region. Thus, we are forced to operate the power-producing steps of these engines in the single-phase vapor region. This is done with the Rankine cycle.

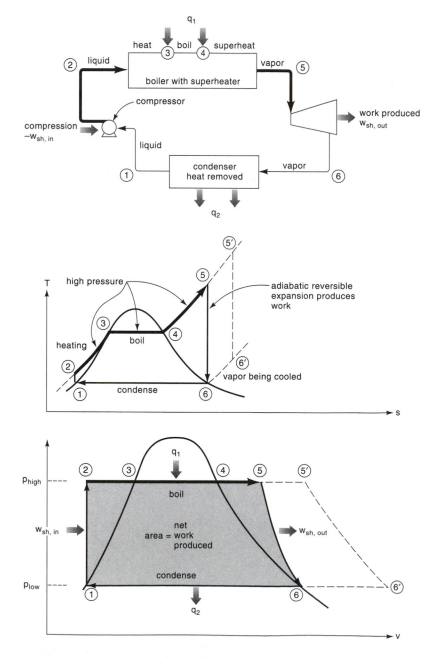

Figure 20–6.

Let us look at various forms of the Rankine cycle. Path 1-2-3-4-5-6-1 of Figure 20–6 represents the ideal Rankine cycle. Assuming negligible kinetic and potential energy contributions, the efficiency of this cycle (as with the Carnot cycle) is found to be

$$\text{efficiency } \eta = \frac{|q_1| - |q_2|}{|q_1|} = \frac{(h_5 - h_2) - (h_6 - h_1)}{h_5 - h_2} \tag{20–3}$$

and with $h_1 \cong h_2$ (to compress a liquid requires very little work) we get

$$\eta = \frac{h_5 - h_6}{h_5 - h_1} \tag{20–4}$$

The efficiency of Rankine cycles can best be compared with Carnot cycles on the T-s diagram, as shown in Figure 20–7.

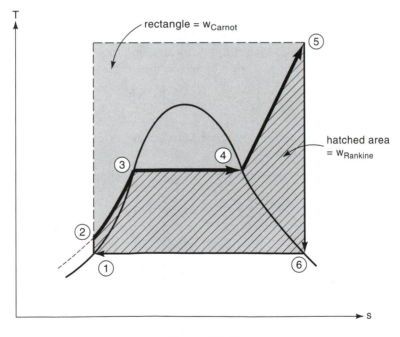

Figure 20–7.

If we are able to use the cycle 1-2-3-4-4', the efficiency would be much closer to the Carnot, (Figure 20–8).

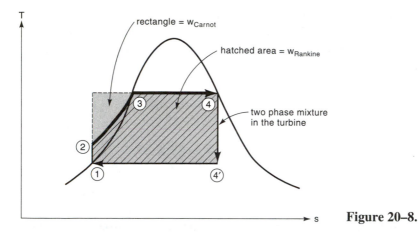

Figure 20–8.

However, as mentioned earlier, there are mechanical problems involved in trying to run a compressor (step 4-4′) with a liquid vapor mixture.

Let us consider some safety considerations. To insure that the expanding vapor does not condense in the turbine and ruin it—in other words, so that point 6 in Figure 20–7 does not fall to the left of where it is shown—the vapor is usually heated somewhat beyond point 5, for example point 5′. In this situation the cycle becomes 1-2-3-4-5′-6′-1, in which case the efficiency becomes (with $h_2 \cong h_1$)

$$\eta = \frac{h_{5'} - h_{6'}}{h_{5'} - h_1} \tag{20–5}$$

But note that doing this lowers the efficiency, as shown in Figure 20–9.

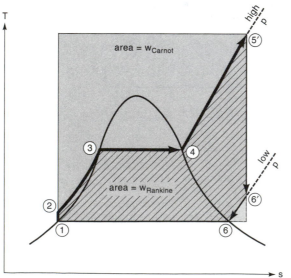

Figure 20–9.

Reheat Rankine Cycles

To approach the Carnot cycle more closely and to avoid the extremely high temperature of point 5′, we go to a reheat cycle. Figure 20–10 shows the simplest of these, the single reheat cycle.

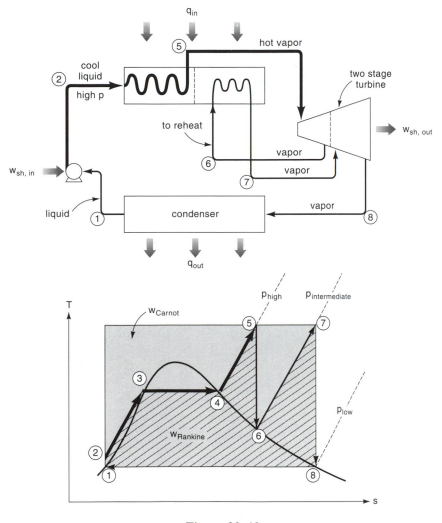

Figure 20–10.

Many variations and alternative cycles have been proposed and are being used, such as

- Multiple reheat cycles (Figure 20–11)
- Regenerative cycles
- Use of feedwater heaters

and so on. We will not go into the details of such systems.

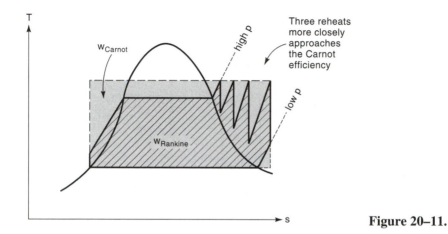

Figure 20–11.

D.   THE RANKINE REFRIGERATION CYCLE

We can closely approximate the Carnot refrigeration cycle discussed in Section C of Chapter 18 if we operate in the two-phase regime for the circulating fluid, shown in Figure 20–12 as the cycle 1'-2'-3-4'→1.

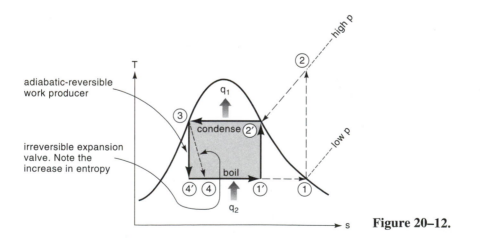

Figure 20–12.

However, it is preferable to compress a pure gas rather than a gas-liquid mixture; thus, we'd like to shift point 1' to point 1. Also, since an adiabatic reversible expansion, 3-4', produces only little work, we can replace this complicated work producer with a simple expansion valve with but a very minor loss in efficiency. These two changes then lead us to the 1-2-3-4-1 cycle, called the Rankine refrigeration cycle, whose behavior is sketched in Figure 20–13.

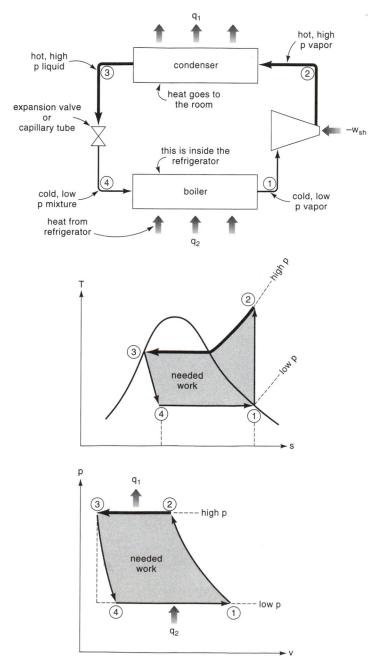

Figure 20–13.

The coefficient of performance of the Rankine refrigerator or air conditioner, from first and second law, is

$$(COP)_{refrig} = \frac{\text{heat removed from refrigerator}}{\text{work input}} = \frac{|q_2|}{|w|} = \frac{h_1 - h_4}{h_2 - h_1} \qquad (20\text{--}6)$$

From the analogous Rankine heat pump we are interested in the heat received, hence

$$COP)_{\text{heat pump}} = \frac{\text{heat pumped into home}}{\text{work input}} = \frac{|q_1|}{|w|} = \frac{h_2 - h_3}{h_2 - h_1} \qquad (20\text{--}7)$$

EXAMPLE 20–1.   A Waste Gobbler

Our engineering college is presently developing a compact power plant specifically designed to use agricultural and forest waste as fuel. The heart of this process is a unique fluidized bed water boiler. In the prototype shown in Figure 20–14, the temperature in the tubes is limited to 300°C while the condenser will operate at 75 kPa.

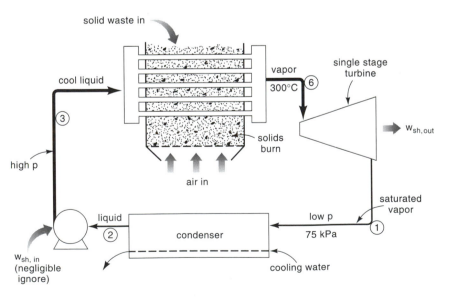

Figure 20–14.

Assuming an ideal Rankine cycle with a saturated power turbine exhaust

(a) On a T-s diagram sketch your cycle.

(b) Recommend a reasonable boiler pressure.

(c) Determine the efficiency of transformation of heat entering the boiler tubes to work.

(d) Find the efficiency of a Carnot cycle operating between these same two temperature limits.

Solution

First let us sketch the T-s diagram and include all the given information on it (Figure 20–15).

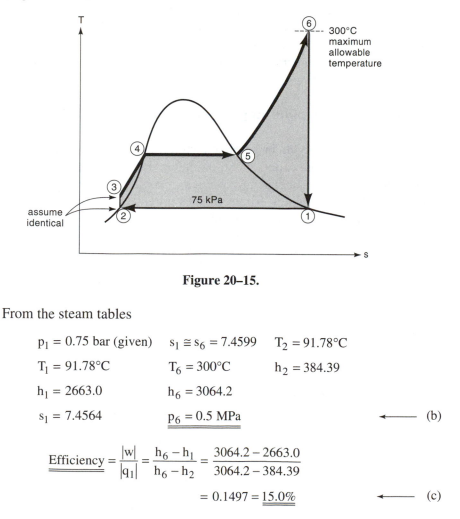

Figure 20–15.

From the steam tables

$$p_1 = 0.75 \text{ bar (given)} \qquad s_1 \cong s_6 = 7.4599 \qquad T_2 = 91.78°C$$

$$T_1 = 91.78°C \qquad T_6 = 300°C \qquad h_2 = 384.39$$

$$h_1 = 2663.0 \qquad h_6 = 3064.2$$

$$s_1 = 7.4564 \qquad \underline{p_6 = 0.5 \text{ MPa}} \qquad\qquad \longleftarrow \text{ (b)}$$

$$\underline{\underline{\text{Efficiency}}} = \frac{|w|}{|q_1|} = \frac{h_6 - h_1}{h_6 - h_2} = \frac{3064.2 - 2663.0}{3064.2 - 384.39}$$

$$= 0.1497 = \underline{15.0\%} \qquad\qquad \longleftarrow \text{ (c)}$$

$$\text{Carnot efficiency} = \frac{T_{hot} - T_{cold}}{T_{hot}} = \frac{(300 + 273) - (91.78 + 273)}{(300 + 273)}$$

$$= 0.363 = \underline{\underline{36.3\%}} \qquad\longleftarrow\qquad (d)$$

E. ONE-PASS GAS ENGINES

Cars, trucks, airplanes both large and small, bulldozers and their cousins that help us reshape our earth's surface are machines designed to extract useful work from the chemical energy bound up in fuels. They do this by various manipulations—heating, cooling, compressing, and expanding gases generated by the fuel. We analyze these changes by approximating them by the ideal gas law.

To illustrate this approach consider the internal combustion engine, of which over 800 million are presently in use worldwide. The working fluids in these engines do not operate in cycles but are analyzed that way. All sorts of engines are of this type: gasoline, diesel, jet, jet turbine, and so on.

1. The Ideal Gasoline Engine—The Otto Cycle

The ordinary piston-cylinder gasoline engine completes a cycle in four strokes of the piston, and is thus called a 4-stroke internal combustion engine. Some small outboard, motorcycle, and chain saw engines complete a cycle in two strokes and are called 2-stroke engines. These engines are analyzed as the air standard Otto cycle, and Figure 20–16 shows the operation of a 4-stroke engine.

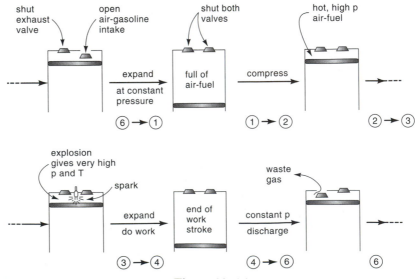

Figure 20–16.

To determine the efficiency of this engine, consider one mole of ideal gas of constant c_v (independent of temperature). Referring to Figures 20–16 and 20–17, which show behavior of the gas in the cylinder, we have,

- Step 1-2
  (adiabatic-reversible compression)
  $$\begin{cases} q & = 0 \\ |w_{in}| & = -\Delta u_{21} \\ \Delta s & = 0 \end{cases}$$

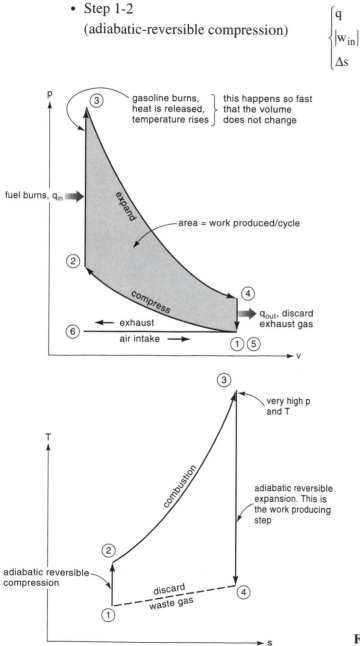

Figure 20–17.

- Step 2-3

    (heat addition at constant volume)

$$\begin{cases} |q_{in}| = c_v(T_3 - T_2) \\ w = 0 \\ \Delta s > 0 \end{cases}$$

- Step 3-4

    (adiabatic reversible expansion)

$$\begin{cases} q_{43} = 0 \\ |w_{out}| = -\Delta u_{43} \\ \Delta s = 0 \end{cases}$$

- Steps 4-5-6-1 can be represented

    by step 4-1 alone

    (constant volume heat removal)

$$\begin{cases} |q_{out}| = c_v(T_4 - T_1) \\ w = 0 \\ \Delta s < 0 \end{cases}$$

Combining all these terms gives the efficiency of the Otto cycle engine

$$\eta = \frac{|q_{in}| - |q_{out}|}{|q_{in}|} = \frac{c_v(T_3 - T_2) - c_v(T_4 - T_1)}{c_v(T_3 - T_2)}$$

and for an ideal gas, noting from Eq. 11–16 that

$$\frac{T_2}{T_1} = \left(\frac{v_1}{v_2}\right)^{k-1} = \left(\frac{v_4}{v_3}\right)^{k-1} = \frac{T_3}{T_4}$$

we find

$$\boxed{\eta = 1 - \frac{T_1}{T_2} = 1 - r_c^{1-k}} \qquad (20\text{--}8)$$

$$\text{compession ratio: } r_c = \frac{v_1}{v_2} = \frac{v_4}{v_3} \qquad (20\text{--}9)$$

The efficiency of a gasoline engine is determined by its compression ratio. The higher the r_c the higher the efficiency. Automobiles have r_c of about 8 or 9. We would like to use higher ratios; however, preignition of the air-gasoline mixture during step 1-2 limits the compression ratio. Additives in present day fuels allow us to use $r_c = 8{\sim}9$. Older cars have $r_c \cong 6$.

This, then, is the air standard gasoline engine. Real gasoline engines approximate this ideal.

2. The Ideal Diesel Engine—The Diesel Cycle

In the gasoline engine the fuel-air mixture is compressed and then burns (explodes) practically instantaneously on action of a spark. So the assumption is made that heat addition is instantaneous.

In the diesel engine air alone is compressed, then diesel fuel is introduced and burns as it mixes with the hot compressed air. This combustion is slower than for a gasoline engine, so the piston moves out during combustion. Thus, we have the behavior shown in Figure 20–18. The corresponding p-v and T-s diagrams are shown in Figure 20–19.

Let us now evaluate the efficiency of the diesel engine. For this, define two ratios as follows

$$\text{the compression ratio}: \quad r_c = \frac{v_1}{v_2} \tag{20-10}$$

$$\text{the expansion ratio}: \quad r_e = \frac{v_4}{v_3} \tag{20-11}$$

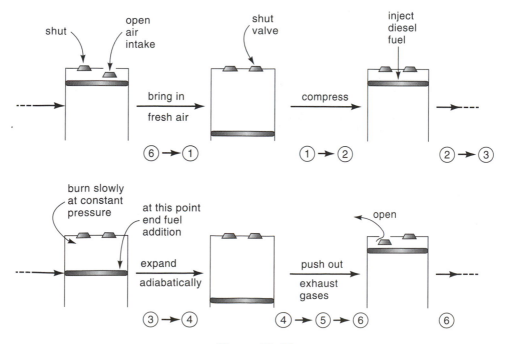

Figure 20–18.

then with ideal gases

$$T_1 = T_2 \left(\frac{1}{r_c} \right)^{k-1}$$

$$T_4 = T_3 \left(\frac{1}{r_e} \right)^{k-1}$$

Combining the above with the efficiency expression gives

$$\eta = \frac{|w|}{|q_{in}|} = \frac{|q_{23}| - |q_{41}|}{|q_{23}|} = \frac{c_p(T_3 - T_2) - c_v(T_4 - T_1)}{c_p(T_3 - T_2)} = 1 - \frac{1}{k} \left(\frac{T_4 - T_1}{T_3 - T_2} \right)$$

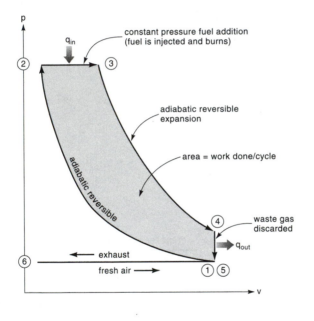

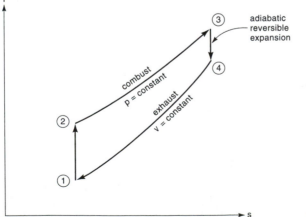

Figure 20–19.

or

$$\eta = 1 - \frac{1}{k} \frac{\left(\dfrac{1}{r_e}\right)^k - \left(\dfrac{1}{r_c}\right)^k}{\dfrac{1}{k_e} - \dfrac{1}{k_c}} \qquad (20\text{-}12)$$

3. Comparison of Gasoline and Diesel Engines

Figure 20–20 compares the Otto and the Diesel cycles for the same and for different compression ratios. It shows that for the same compression ratio the Diesel has a lower efficiency. However, the Diesel can operate at much higher compression ratios, up to 20, in which case it is much more efficient than the Otto engine.

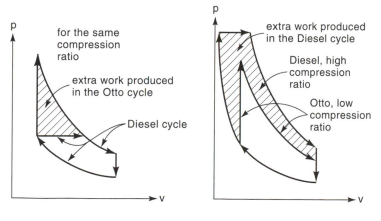

Figure 20–20.

In real engines the various steps are not as distinct as shown in Figures 20–17 and 20–20, because the adiabatic steps are not truly adiabatic, friction occurs, and so on. Despite this, the approach shown is the way engineers approach the operation of these engines.

4. Combustion-Gas Turbine (Brayton or Joule Cycle)

A turbine in a stationary power plant (Rankine cycle) is more efficient (in terms of friction) than the reciprocating engine. On the other hand, internal combustion has advantages over an external heat source engine (less complicated). The combustion turbine tries to combine the advantages of these two types of engines.

Figure 20–21 sketches the main features of a combustion gas turbine and then shows the ideal p-v and T-s diagram per mole of air passing through this engine. Referring to these figures, the efficiency of this engine is

$$\eta = \frac{|w_{CD}| - |w_{AB}|}{|q_{BC}|} = \frac{|q_{BC}| - |q_{AD}|}{|q_{BC}|}$$

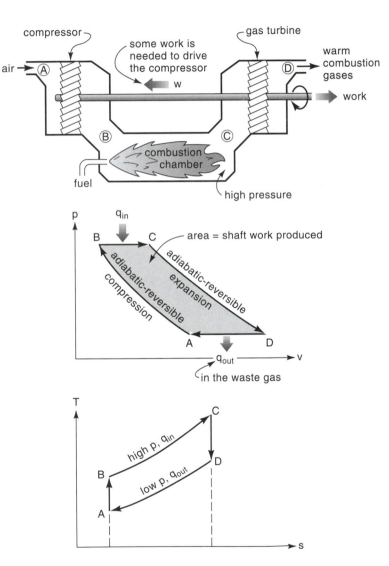

Figure 20–21.

For an ideal gas

$$\eta = \frac{c_p(T_C - T_B) + c_p(T_A - T_D)}{c_p(T_C - T_B)}$$

and for adiabatic reversible expansion and compression between $p_B = p_C$ and $p_A = p_D$

$$\frac{T_B}{T_A} = \left(\frac{p_B}{p_A}\right)^{(k-1)/k} = \left(\frac{p_C}{p_D}\right)^{(k-1)/k} = \frac{T_C}{T_D}$$

Combining the above two expressions gives

$$\boxed{\eta = 1 - \frac{T_A}{T_B} = 1 - \left(\frac{p_A}{p_B}\right)^{(k-1)/k}} \qquad (20\text{--}13)$$

Comments.   A large fraction of output work is needed to run the compressor (~60%) as opposed to the Rankine cycle, which compresses a liquid (~1%). So the big problem is to design an efficient compressor. Today compressor efficiencies are about 80%.

Equation 20–13 tells that the higher the combustion temperature the more efficient is the engine. This temperature is limited by the strength of the metal turbine blades. Ceramic turbine blades retain their strength at much higher temperatures and are being introduced in the latest turbines.

5.   Jet Propulsion Engines—Brayton or Joule Cycle

Here the shaft work of the combustion gas turbine is replaced by kinetic energy of the exhaust gases. There are several types of jet engines.

6.   Turbojet Engines

These engines employ gas turbines and use the kinetic energy of the exhaust gases. This is sketched in Figure 20–22.

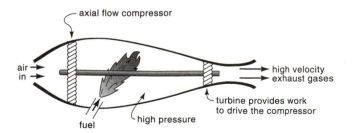

Figure 20–22.

7.   Ramjet Engines

If air enters the engine at high enough velocity, the pressure builds up high enough so that we can dispense with both compressor and turbine (Figure 20–23). This engine gives good efficiency at supersonic velocities; otherwise, it is not useful at all.

Figure 20–23.

8.   Force of Propulsion

Let us briefly sketch the concepts used to analyze these jet engines. From Newton's third law the force of propulsion, called *the thrust*, is given by

$$F = \frac{(\dot{m}u)_{\text{leaving}} - (\dot{m}u)_{\text{entering}}}{g_c} \tag{20–14}$$

and neglecting the mass of fuel, since it is small compared to the mass of air (with its nitrogen) passing through the engine, we write

$$F = \frac{\dot{m}_{\text{air}}(u_{\text{leaving}} - u_{\text{entering}})}{g_c} = \frac{\dot{m}\,\Delta u}{g_c} \tag{20–15}$$

Now from first law we have

$$\Delta h + \frac{\Delta u^2}{2g_c} + \Delta e_p = q - w_{\text{sh}}$$

or

$$c_p \Delta T + \frac{\Delta u^2}{2g_c} = 0$$

This shows that the temperature of the gases measures the gas velocity, or

$$F \propto \left(\sqrt{T_{\text{leaving}}} - \sqrt{T_{\text{entering}}} \right) \tag{20–16}$$

The meaning of this expression is that the larger the Δh_r of the fuel, the larger is the thrust of the engine.

9. Rocket Engines (Figure 20–24)

Here we have behavior similar to the jet engine except that there is no compressor, and the working fluid is carried with the rocket. Such engines are ideal for outer space applications.

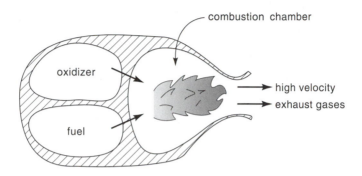

Figure 20–24.

Since there is no air intake, set $u_1 = 0$ and Eq. 20–14 becomes

$$F = \frac{\dot{m}_{\text{leaving}} u_{\text{leaving}}}{g_c} \tag{20–17}$$

The important measure of efficiency of rocket engines is the *specific impulse*. This is defined by

$$SI = \frac{(\text{thrust}) (\text{time})}{\begin{array}{c}\text{unit mass of}\\ \text{fuel and}\\ \text{oxidizer used}\end{array}} = \frac{F}{\dot{m}} = \frac{u_{\text{leaving}}}{g_c} \quad \left[\frac{N \cdot s}{kg}\right] \tag{20–18}$$

So the velocity, or temperature, of the leaving gases is all that matters in determining the specific impulse, maximum power, efficiency, and so on. We leave these considerations to the specialist.

PROBLEMS

1. Steam at 8 MPa and 800°C enters a turbine and expands adiabatically to 3 bar and 300°C. What is the turbine efficiency?

2. An ideal steam power plant operates on the Rankine cycle under the following conditions

    Boiler pressure: 20 bar
    Condenser pressure: 70.14 kPa
    Turbine outlet has 10°C of superheat

    Find the efficiency of the cycle. Please neglect the compression work in your calculations.

3. The geothermal wells of problem 19–20 produce dirty hot water at 200°C. This pressurized liquid goes to a heat exchanger to heat pure water to saturated steam at 180°C. This steam is part of a Rankine cycle whose condenser is at 20 kPa and whose turbine is 100% efficient.

    (a) Sketch the T-s diagram of this Rankine cycle and number all its pertinent points.

    (b) Determine the overall efficiency of this power cycle (ignore pumping power).

    (c) What is the quality (% vapor) of the exit from the turbine?

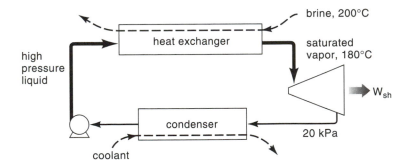

4. Repeat the previous problem with one change: Because of condensation of water in the turbine, it is only 50% efficient when compared with the adiabatic reversible ideal.

5. In its refrigeration plant at Albany, Oregon, Freeze-Rite uses a vapor compression cycle with HFC-134a as the working fluid. Sneaking into the plant and looking at the pressure gauges, I note that one part of their cycle operates at about 50 kPa, the other part of their cycle operates at 1.8 MPa. I also note that an expansion value is present as part of their system. From the above information, could you please determine the coefficient of performance of their refrigeration plant if everything operated ideally?

6. Repeat Example 20–1 with one change: We will use a reheat cycle with a two-stage turbine in place of the one-stage turbine. This modification is shown in the sketch below.

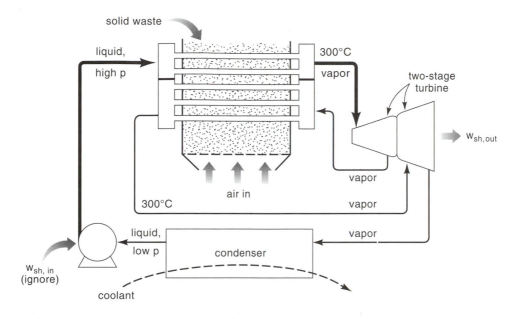

7. For the car I plan to buy I have a choice of two engines: the first has a compression ratio of 8.5 and will burn regular gasoline, the second has a compression ratio of 10.5 and must burn the super grade of gasoline. Assume that the cars are identical except for this fact (same weight, etc.) and that these two gasolines are equivalent as far as energy per gallon of fuel. From the normal outrageous prices of gasoline today (36.9¢/liter for regular and 49.9¢/liter for super), which automobile would be more economical to drive?

8. For the same compression ratios, rate the Otto, Diesel and Brayton cycles.

CHAPTER 21
PHASE EQUILIBRIUM

Up to now we have dealt with single-component systems such as water as liquid and steam, HFC-134a as liquid and gas, and gases of all types. However, chemists, biologists, and chemical engineers often have to deal with mixtures of chemicals, separating them, mixing them, reacting them.

We now extend our single substance treatment to multichemical mixtures. In this chapter we study *phase equilibrium* to tell how the individual components of a mixture distribute themselves between gas and liquid when the two phases are in equilibrium with each other. Later we study *chemical equilibrium*, in which thermo tells what fraction of reactants could in principle react to produce product.

A. MISCIBLE MIXTURES

Here we want to know how two or more chemical components distribute themselves in a two-phase system. In general the treatment of this problem is rather involved, requiring defining concepts of fugacity, activity coefficients, and chemical potentials. We can bypass all of this by treating the simplest of situations, where the gas is an *ideal mixture* and where the liquid behaves as an *ideal solution*. Consider this now.

1. The Ideal Gas Mixture

As Chapter 11 showed, an ideal gas mixture of A and B follows Dalton's Law, or

$$\left.\begin{array}{l} p_A = \pi \, y_A \\ p_B = \pi \, y_B \end{array}\right\} \quad \text{and} \quad \begin{array}{l} p_A + p_B = \pi \\ y_A + y_B = 1 \end{array} \tag{21-1}$$

where y_A and y_B represent the mole fraction of A and of B in the gas and π is the total pressure of the gas.

2.   The Ideal Solution

Consider a gas-liquid mixture of A and B at equilibrium at T and p. Figure 21–1 shows the symbols representing this situation.

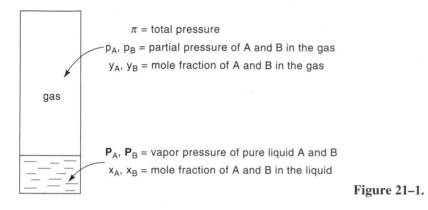

π = total pressure

p_A, p_B = partial pressure of A and B in the gas

y_A, y_B = mole fraction of A and B in the gas

gas

$\mathbf{P_A}$, $\mathbf{P_B}$ = vapor pressure of pure liquid A and B

x_A, x_B = mole fraction of A and B in the liquid

Figure 21–1.

Now an ideal solution is one that follows Raoult's Law, or

$$\left. \begin{array}{l} p_A = \mathbf{P_A} x_A \\ p_B = \mathbf{P_B} x_B \end{array} \right\} \quad \text{and} \quad x_A + x_B = 1 \tag{21-2}$$

The idea behind the assumption of ideal gas and ideal solution is that a molecule of A treats its surrounding molecules of A and B alike—no extra repulsion, no extra attraction, no preference for either A or B, no discrimination. So if you mix one volume of A with one volume of B you would get exactly two volumes of mixture, not less (if A and B attract each other), and not more (if A and B repel each other).

This is the normal situation for molecules that are similar in structure such as mixtures of methanol-ethanol-propanol, or hexane-heptane-octane, or benzene-toluene, and so on. Dissimilar substances often do not behave like ideal solutions; if they are very dissimilar they often won't form a solution at all, such as oil and water.

The Cox chart of Figure 21–2 relates the vapor pressure of pure liquids with temperature. For most materials this relationship can be reasonably represented by a straight line. Evaluate the pressure at two temperatures—for example, the critical and normal boiling points—and join to give the whole range of vapor pressures.

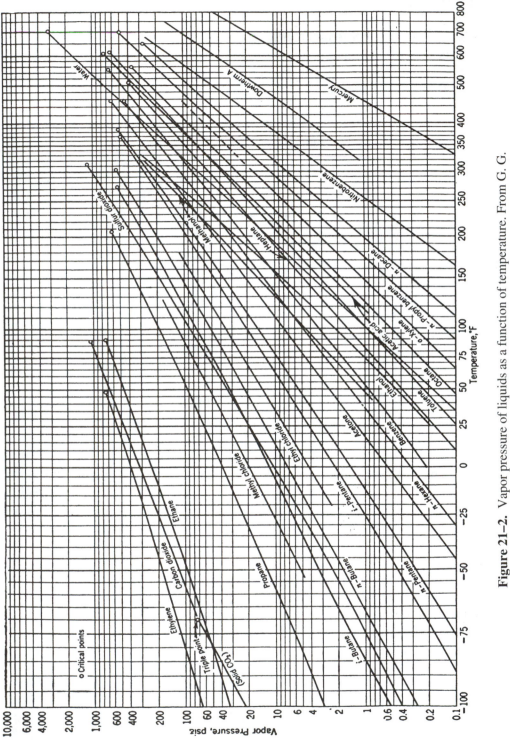

Figure 21–2. Vapor pressure of liquids as a function of temperature. From G. G. Brown et al. *Unit Operations* (Courtesy of John Wiley and Sons, New York: 1950) p. 583.

3.   The Phase Equilibrium Constant, K

Combining Eqs. 21–1 and 21–2 gives

$$\left.\begin{aligned} p_A = \pi\, y_A = \mathbf{P}_A x_A \\ p_B = \pi\, y_B = \mathbf{P}_B x_B \end{aligned}\right\} \tag{21-3}$$

The phase equilibrium constant K is a convenient supplemental variable defined as

$$\left.\begin{aligned} K_A = \frac{y_A}{x_A} = \frac{\mathbf{P}_A}{\pi} \\[2em] K_B = \frac{y_B}{x_B} = \frac{\mathbf{P}_B}{\pi} \end{aligned}\quad\right\}\quad \text{\textit{only for ideal solutions}} \tag{21-4}$$

Values of K for various materials are shown in Figure 21–3. For a material whose liquid has a very low vapor pressure, $K \to 0$, while for a material whose liquid has a very high vapor pressure, say hydrogen or oxygen at room temperature, $K \to \infty$.

What we say for two-component system can be extended directly to multi-component systems.

4.   Strategy for Solving Phase Equilibrium Problems

Consider a feed of F moles of A and B having mole fractions z_A and z_B, which splits into a gas-liquid mixture at equilibrium. Figure 21–4 shows the symbols describing this split.

Material balances for the systems of Figure 21–4 give

$$F = G + L \quad \left.\begin{aligned} z_A F = x_A L + y_A G \\ z_B F = x_B L + y_B G \end{aligned}\right\} \quad \left.\begin{aligned} x_A + x_B = 1 \\ y_A + y_B = 1 \end{aligned}\right\} \quad \left.\begin{aligned} y_A = K_A x_A \\ y_B = K_B x_B \end{aligned}\right\} \tag{21-5}$$

A useful combination of variables of Eq. 21–5 gives, for component A,

$$x_A = \frac{z_A F}{L + K_A(1-L)}, \qquad y_A = \frac{z_A F}{G + \dfrac{(1-G)}{K_A}} \tag{21-6}$$

Similar equations hold for component B. We will see how to use these equations to solve phase equilibrium problems.

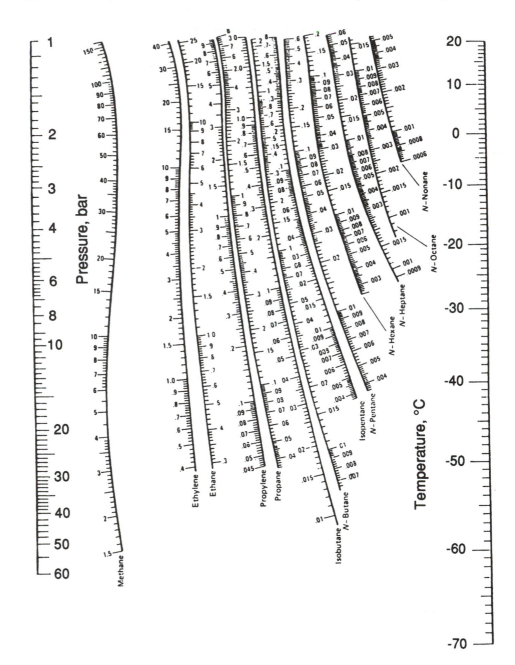

Figure 21–3a. K-values in light-hydrocarbon systems; low-temperature range. Adapted from D. B. Dady Burjor. *Chem. Eng. Prog.*, pp. 85, 86, April 1978 with permission of AICHE.

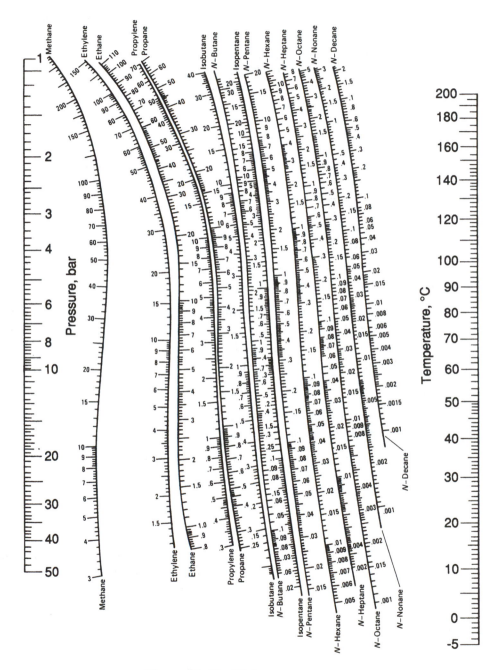

Figure 21–3b. High-temperature range.

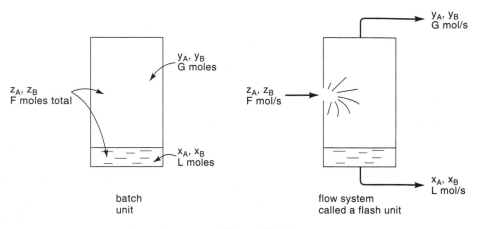

Figure 21–4.

EXAMPLE 21–1. The flash separator

A 40 mol% isobutane 60 mol% normal pentane mixture flows into a chamber, flashes at 49°C and 3.2 bar, and leaves as two streams, one gas, the other liquid. Find how much gas and liquid leave per mole of entering feed, and find the composition of these leaving streams.

Solution

Let sub 1 refer to i-butane and sub 2 refer to n-pentane. Then, for one mole of feed mixture write what is known.

$$F = 1 \text{ mole}, \ z_1 = 0.4, \ z_2 = 0.6, \text{ and from Fig. 3 } K_1 = 2, \ K_2 = 0.5$$

So Eq. 21–6 becomes

$$x_1 = \frac{0.4 \, (1)}{L + 2 \, (1 - L)} \qquad x_2 = \frac{0.6 \, (1)}{L + 0.5 \, (1 - L)} \qquad \text{(i)}$$

In general, at this point trial and error methods are used to solve phase equilibrium problems. Let us do this here, guessing values for L until $x_1 + x_2 = 1$.

Guess	L = 0.5 from eq i	L = 0.1	L = 0.9	L = 0.8
	x_i	x_i	x_i	x_i
i - C_4	0.27	0.21	0.36	0.333
n - C_5	0.80	1.09	0.63	0.667
$\Sigma x_i =$	1.07	1.30	0.99	1.000

correct

or

$$L = 0.8 \quad \begin{cases} x_1 = 0.33 \quad \text{———} \\ x_2 = 0.67 \end{cases}$$

and with $y_i = K_i x_i$

$$G = 0.2 \quad \begin{cases} y_1 = 0.67 \quad \text{←———} \\ y_2 = 0.33 \end{cases}$$

EXAMPLE 21–2.   University engineering goes practical

Our university's engineering school decided that it was time to do something about the energy crisis—enough talk—so we set up our own oil drilling research project (Figure 21–5). We took off directly under the Dean's desk (can you think of a better place?) and drilled down, down, down, past the gas mains, sewers, and so on, heading towards the center of the earth and the Persian oil wells on the other side. However, we were only partly successful because our well only evolved high pressure gas with negligible liquid. As energy experts we are convinced that this is only a gas cap over a giant pool of oil, so we plan to drill further until we reach the liquid.

To be prepared for processing the liquid we must know what to expect, a gooey mass or what. Assuming the gas to be in equilibrium with the liquid below it, estimate the composition of the liquid. I suppose you can estimate the temperature way down there to be 50°C.

Data

Composition of the gas evolved

$$CH_4 - 80\%, \ C_2H_6 - 10\%, \ C_3H_8 - 10\%$$

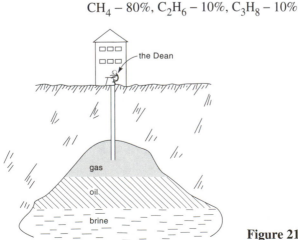

Figure 21–5.

Solution

For one mole of gas let us see what is known

$$y_i = K_i x_i \quad \text{or} \quad x_i = \frac{y_i}{K_i} \tag{i}$$

Now guess the pressure, find K_i from Figure 21–3, then evaluate x_i, and solve by trial and error for the pressure that gives $\Sigma x_i = 1$.

Given		Guess: $\pi = 10$ bar		Guess: $\pi = 50$ bar	
	y_i		from eq i		
		K_i	x_i	K_i	x_i
C_1	0.8	19.4	0.0412	3.9	0.2051
C_2	0.1	4.5	0.0222	1.28	0.0781
C_3	0.1	1.67	0.0599	0.046	0.2174
$\Sigma x_i =$			0.1233		0.5006

     ↖ much too low   ↖ still much too low

Even at the highest pressure on the graph of Figure 21–3 Σx_i is too low, nowhere near $\Sigma x_i = 1$. Since the graph ends near the critical point of these materials, this means that the mixture is at a pressure above the critical point, and this in turn means that the gas reaching the surface and the fluid down deep have the same composition.

  Therefore, the reservoir consists of

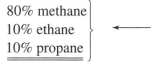

$$\left.\begin{array}{l} \underline{80\% \text{ methane}} \\ \underline{10\% \text{ ethane}} \\ \underline{10\% \text{ propane}} \end{array}\right\} \longleftarrow$$

B. IMMISCIBLE MIXTURES

When components A and B are completely immiscible as liquids and form separate liquid phases, they are still miscible as gases. Thus, we can look at this mixture as shown in Figure 21–6. The equation which represents this situation is

$$\pi = p_A + p_B = P_A x_A + P_B x_B = P_A + P_B$$

    $\underset{\text{pure A in its phase; } \therefore\, x_A = 1}{\underline{}}$

$$\tag{21-7}$$

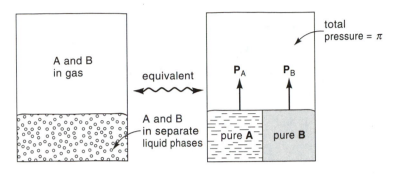

Figure 21–6.

C. COMPOUND SYSTEMS

For compound systems where A and B form an ideal solution that is immiscible with C, we have the situation sketched in Figure 21–7. The equations that express the equilibrium between the three phases are

$$\pi = p_A + p_B + p_C = \mathbf{P}_A x_A + \mathbf{P}_B x_B + \mathbf{P}_C x_C \qquad (21\text{–}8)$$

$$x_A + x_B = 1, \qquad x_C = 1$$

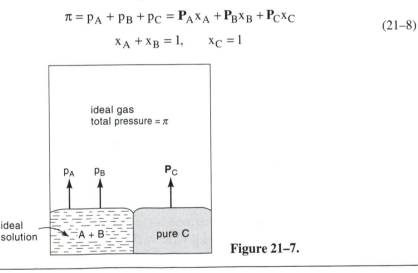

Figure 21–7.

EXAMPLE 21–3.   Immiscible equilibrium

Find the composition of vapor in equilibrium with a liquid mixture of water, hexane, and heptane at 92°C and 219 kPa. Note that water is completely immiscible with the two organics, which form an ideal solution.

Solution

Let sub 1 refer to hexane, sub 2 refer to heptane and sub w refer to water. Then
from the water-steam tables or Figure 21–2: $p_w = 73$ kPa
from Figure 21–3: $K_1 = 0.9$, $K_2 = 0.4$

From Eqs. 21–3 and 21–4

$$p_1 = \pi \, y_1 = \pi \, K_1 x_1 = 219 \, (0.9) \, x_1 = 197.1 \, x_1 \text{ kPa}$$

$$p_2 = \pi \, y_2 = \pi \, K_2 x_2 = 219 \, (0.4) \, x_2 = 87.6 \, x_2 \text{ kPa}$$

$$p_w = 73 \text{ kPa}$$

But $x_1 + x_2 = 1$, and

$$p_1 + p_2 = \pi - p_w = 219 - 73 = 146 = 197.1 \, x_1 + 87.6(1 - x_1)$$

from which

$$x_1 = \frac{146 - 87.6}{197.1 - 87.6} = 0.533$$

$$x_2 = 0.467$$

So for the gas phase

$$\left. \begin{array}{l} p_{\text{hexane}} = \pi \, K \, x_1 = 219 \, (0.9) \, (0.533) = 105.12 \text{ kPa} \\[2mm] p_{\text{heptane}} = 219 \, (0.4) \, (0.467) = 40.88 \text{ kPa} \\[2mm] p_{\text{water}} = 73 \text{ kPa} \end{array} \right\} \longleftarrow$$

PROBLEMS

1. A 50% normal pentane 50% normal octane mixture (mole basis) flows into a chamber, flashes at 146°C and 3.6 bar, and leaves as two streams, one gas, the other liquid. Find the flow rate and composition of the two streams.

2. A stream of 40 mol% ethane-60 mol% propane flashes into a chamber kept at 10 bar. Of the entering stream 20 mol% leaves as vapor, 80 mol% leaves as liquid. Find the composition of the vapor and its temperature.

3. A 50-50 (mole basis) mixture of ethylene and propylene is to be introduced into a tank that is to be kept at 5 bar. At what temperature would it be 50 mol% liquid, 50 mol% vapor?

4. A 50-50 mixture of n-pentane, n-hexane (mole basis) is subjected to a flash distillation at 100°C. Of the total feed 30 mol% leaves the flash unit as vapor. Calculate the compositions of the distillate stream and the residual liquid stream.

5. An equimolar mixture of ethylene and propylene is 50 mol% vapor, 50

mol% liquid at 20°C. What is the vapor pressure of the mixture and the vapor composition?

6. A stream of 20% hydrogen, 30% isobutane, and 50% normal pentane flashes into a chamber kept at 3.2 bar and 49°C and leaves as two streams, one gas, the other liquid. Find the flow rate and composition of the gas stream.

7. A mixture of 20 mol% ethane, 50 mol% propane and 30 mol% duodecane ($C_{12}H_{26}$) is flashed at 12 bar and 32°C. Find the composition of the vapor.

8. A stream of 30% hydrogen, 20% isobutane, 40% normal pentane, and 10% decane ($C_{10}H_{22}$) flashes into a chamber kept at 3.2 bar and 49°C and leaves as two streams, one gas, the other liquid. Find the flow rate and compositions of the two streams.

9. A swoosh of water, heptane, and octane (liquid organics, immiscible with water) plus a bit of vapor are being pumped through a pipeline at 105°C and 2 bar. Find the molar % composition of the vapor.

10. As a teaser, and for a bit of fresh air, consider the sketch below of a rowboat on a little pond. A big heavy anchor is thrown overboard from the rowboat. What does this do to the water level of the pond? Does it rise, fall or stay unchanged?

> Beware: According to J. Walker, *The Flying Circus of Physics* (Wiley, 1975), this problem was asked of George Gamow, Robert Oppenheimer, and Felix Bloch, all excellent physicists, and to their embarrassment they all answered incorrectly.

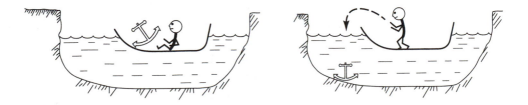

CHAPTER 22

MEMBRANES, FREE ENERGY, AND WORK FUNCTIONS

A. FREE ENERGY AND THE WORK FUNCTION

A Batch of Material

When a batch of material goes from state 1 to state 2, the work and heat interactions with the surroundings are related to the energy change of the system according to

$$Q_{1\to2} - W_{1\to2} = \Delta U + \Delta E_p + \Delta E_k \qquad \text{(3–2) or (22–1)}$$

When all the mechanical energy changes occur reversibly (no frictional effects) and both the system and surroundings are at the same temperature T_{syst} throughout, then Eq. 15–3 gives $Q_{rev} = T_{syst}\Delta S$, so that

$$W_{1\to2} = -(\Delta U - T_{syst}\Delta S + \Delta E_p + \Delta E_k) \qquad (22–2)$$

This expression gives the maximum useful work that can be obtained from the system as it goes from state 1 to state 2 when both the system and surroundings are at T_{syst}.

Note that $W_{1\to2}$ is not the exergy $W_{ex,1\to2}$ (see Eq. 19–7) because $W_{1\to2}$ includes pV work as the system expands, but does not account for the extra work that can be produced when heat flows between system and surroundings in the more general situation where $T_{syst} \neq T_{surr}$.

When a system starts in state 1, reversibly passes through state 2, and eventually ends up at equilibrium, state 3, more and more work can be extracted. We show this graphically in Figure 22–1.

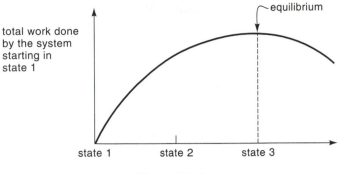

Figure 22–1.

Thus, from Eq. 22–2, Figure 22–2 shows the results graphically.

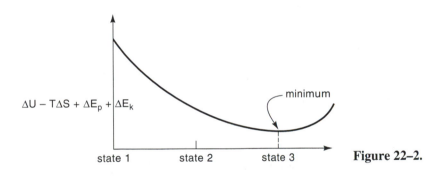

Figure 22–2.

At equilibrium $E_k = 0$, so for any change to a state away from equilibrium

$$\frac{dW}{d \text{ (any change)}} < 0 \qquad (22\text{–}3)$$

At equilibrium $E_k = 0$, so for any two points in a system at equilibrium

$$\boxed{\begin{array}{c} W = 0 \\ \Delta U - T\Delta S + \Delta E_p = 0 \end{array}} \qquad (22\text{–}4)$$

$$\textit{equilib, } T_{syst} = T_{surr}$$

As we will later see, this expression tells that when a column of a gas mixture is at equilibrium, then both the composition ratio and the pressure will differ from bottom to top. Also, if the ocean surface contains 3% salt, then down deep the percentage of salt will be quite different.

Steady State Flowing Material

The first law gives the energy interactions between upstream point 1 and down-stream point 2 as

$$Q_{1\rightarrow 2} - W_{sh,1\rightarrow 2} = \Delta H + \Delta E_p + \Delta E_k \qquad \text{(13–3) or (22–5)}$$

If the material flows isothermally and frictionlessly, then $Q_{rev,1\rightarrow 2} = T_{syst}\Delta S$ (see Eq. 15–3), so that Eq. 22–5 becomes

$$W_{sh,1\rightarrow 2} = -(\Delta H - T\Delta S + \Delta E_p + \Delta E_k) \qquad (22\text{–}6)$$

In this analysis, we assume that the system and surroundings are both at T_{syst}.

Now this combination of H, T_{syst}, and S appears again and again in advanced thermo, so let us give it a single symbol G, called the Gibbs free energy, thus

$$G = H - TS \qquad (22\text{–}7)$$

so at constant temperature

$$\Delta G = \Delta H - T\Delta S \qquad (22\text{–}8)$$

Combining Eqs. 22–6 and 22–8 gives

$$W_{sh,1\rightarrow 2} = -(\Delta G + \Delta E_p + \Delta E_k) \qquad (22\text{–}9)$$

Again, as with batch systems, a spontaneous approach to equilibrium can produce work, at equilibrium $E_k = 0$, so for any two points in the flow system at equilibrium

$$\boxed{W_{sh} = 0 \quad \text{and} \quad \Delta G + \Delta E_p = 0} \qquad (22\text{–}10)$$

Also, in going from any state to equilibrium, the amount of work produced is maximized, and so at equilibrium G plus E_p is minimized.

Ideal gas

For an ideal gas undergoing an isothermal process $\Delta H = 0$, $\Delta U = 0$, so for both batch and flow operations

$$W_{sh,1\rightarrow 2} = -T\Delta S + \Delta E_p + \Delta E_k$$

With Eqs. 5–2, 6–2 and 16–4 this expression becomes

$$W_{sh,1\rightarrow 2} = -nRT \; \ell n \; \frac{p_1}{p_2} + \frac{mg(z_2 - z_1)}{g_c} + \frac{m\left(v_2^2 - v_1^2\right)}{2\,g_c}$$

At equilibrium $E_k = 0$, so we have

$$-nRT \ln \frac{p_1}{p_2} + \frac{mg(z_2 - z_1)}{g_c} =$$

or

$$\boxed{\frac{p_1}{p_2} = e^{\frac{\overline{mw}\, g(z_2 - z_1)}{g_c RT}}} \qquad (22\text{–}11)$$

ideal gas, const T, equilibrium

EXAMPLE 22–1. Storage of the world's densest gas

Since the beginning of the atmoic age, the United States has stored uranium as gaseous uranium hexafluoride (UF_6, $\overline{mw} = 0.352$ kg/mol) in abandoned Texas oil wells. This reserve was forgotten, and now fifty years later, we accidentally came across one of those storage wells on a Texas armadillo ranch (Figure 22–3). We sampled the gas at the top of the 2 km deep tank and found it to be pure UF_6 at 2 bar and 300 K. What is the pressure at the bottom of the storage tank? Assume ideal gas behavior.

Solution

After forty years we can assume equilibrium through-out the tank. So, for an isothermal pressure change from bottom (point 1) to top (point 2), Eq. 22–11 gives

$$p_1 = p_2 e^{\frac{\overline{mw}\, g(z_2 - z_1)}{g_c RT}}$$

$$= (2 \text{ bar})\, e^{\frac{0.352\,(9.8)\,(2000)}{1\,(8.314)\,(300)}} = \underline{31.8 \text{ bar}} \longleftarrow$$

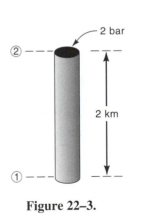

Figure 22–3.

B. SEMIPERMEABLE MEMBRANES

Given either a gas or a liquid mixture of components A and B, we define a semipermeable membrane as one that only allows free passage across the membrane of one of the two components while being completely impervious to the other (Figure 22–4).

In practice such membranes are used

- For water desalination, to produce fresh water from salty or brackish water
- To extract hydrogen from gas mixtures
- To purify blood of dialysis patients and thereby allow hundreds of thousands of people with failed kidneys to live

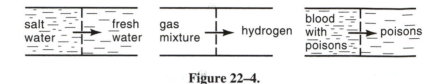

Figure 22–4.

Conceptually, the ideal semipermeable membrane is useful in thermo for developing the basic equations for phase and chemical equilibrium. The examples that follow illustrate this.

EXAMPLE 22–2.   Semipermeable membranes for gases

If it is stored in spherical chambers, radioactive uranium can become critical (have a temperature runaway) and explode. So, in the 1950s safe storage farms were created that consisted of long, small diameter tubes drilled into the ground. These were then filled with a mixture of hydrogen and uranium hexafluoride gas.

These storage tanks were left sealed and untouched for decades so we can assume that their contents have come to equilibrium. One was recently uncapped and was found to consist of a 50-50 mol% mixture at 1 bar at 300 K at the top (point 2). What is the pressure and the composition ratio at the bottom, 2 km below the surface (point 1)? Assume ideal gas behavior.

Solution

To solve this problem we have to introduce the concept of an ideal semipermeable membrane. Also let symbols A and B represent the two components

$$A = H_2, B = UF_6$$

Now refer to Figure 22–5.

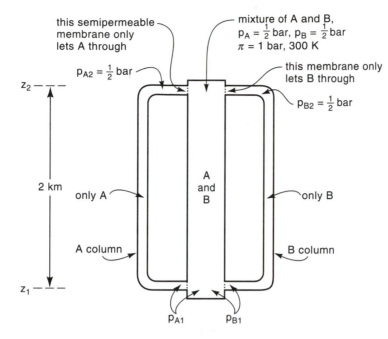

Figure 22–5.

The top and bottom of column A (the pure hydrogen column) are in equilibrium so as with Example 22–1 we apply Eq. 22–11. Thus

$$p_{A1} = p_{A2}e^{\frac{\overline{mw}_A\, g(z_2 - z_1)}{g_c RT}}$$

$$= (0.5\text{ bar})e^{\frac{0.002\,(9.8)\,(2000)}{1\,(8.314)\,(300)}} = 0.508\text{ bar}$$

For column B (the pure UF_6 column), similarly, with data from Example 22–1

$$p_{B1} = (0.5\text{ bar})e^{\frac{0.352\,(9.8)\,(2000)}{1\,(8.314)\,(300)}} = 7.95\text{ bar}$$

Therefore, at the bottom of the storage tank

$$\text{total pressure} = 0.508 + 7.95 = 8.46\text{ bar}$$

$$\left.\begin{array}{l} H_2 = \dfrac{0.508}{0.508 + 7.95} \times 100 = \underline{\underline{6.0\text{ mol\%}}} \\[3mm] UF_6 = \dfrac{7.95}{0.508 + 7.95} \times 100 = \underline{\underline{94.0\text{ mol\%}}} \end{array}\right\} \quad \longleftarrow$$

C. ΔP ACROSS A LIQUID MEMBRANE AT EQUILIBRIUM

For a gaseous mixture at equilibrium the partial pressure of a permeable component is the same on both sides of an ideal membrane and analysis is straightforward (see Example 22–2). For a system involving liquids and vapors, things are more complicated and we have to introduce the concept of ideal solution and ideal gas mixtures. We do this as follows.

Suppose we have a liquid mixture of A and B with mole fraction x_A and x_B on the left side of a membrane that is permeable to A alone, pure A on the right side. Let B have a negligible vapor pressure. Then we have the situation sketched in Figure 22–6. This sketch shows that the vapor pressure on the left is lower than on the right, so component A will flow through the membrane as well as over the top from right to left. This does not represent an equilibrium condition.

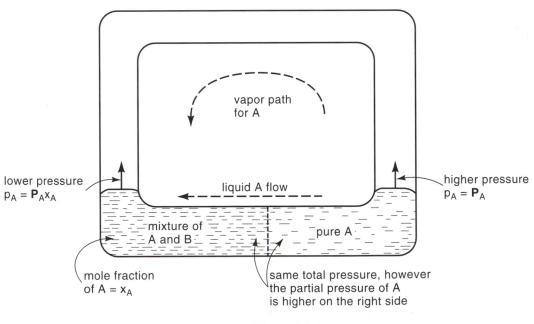

Figure 22–6.

Equilibrium would represent a situation where no permeable component A would flow from one side to the other, and this can be attained if the liquid level is different on the two sides, as shown in Figure 22–7, such that p_A at z_2 is the same on both arms of the loop.

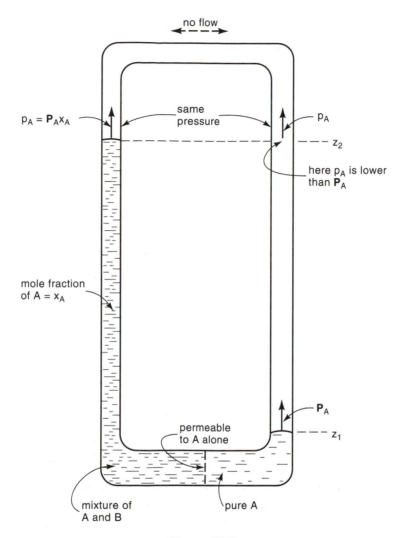

Figure 22–7.

Let us evaluate $\Delta z = z_2 - z_1$

For the pure A side at z_1: $\quad p_A = \mathbf{P}_A$

For the pure A side at z_2: $\quad p_A = \mathbf{P}_A e^{-\dfrac{\overline{mw}_A \; g(z_2 - z_1)}{g_c RT}}$ (22–12)
(from Eq. 22–11)

For the A - B side at z_2: $\quad p_A = \mathbf{P}_A x_A$ (22–13)

Combining Eqs. 22–12 and 22–13 relates the mole fraction of A in solution with the difference in height, as follows

$$x_A = e^{-\dfrac{\overline{mw_A}\, g\, \Delta z}{g_c RT}}$$

or

$$\Delta z = \frac{g_c RT}{mw_A\, g}\; \ell n \frac{1}{x_A} \tag{22–14}$$

Now, to find the pressure on the mixture side at z_1, write the first law between z_1 and z_2,

$$\cancel{\Delta U} + \Delta E_p + \cancel{\Delta E_k} = \cancel{Q} - \cancel{W}_{sh} - W_{pV} \qquad \text{from (7–4)}$$

Replacing values, as in Eq. 13–15, we get per kilogram of mixture

$$\frac{g(z_2 - z_1)}{g_c} = -\frac{(p_2 - p_1)}{\rho_{mixture}} \tag{22–15}$$

And combining with Eq. 22–8

$$\overbrace{(p_1 - p_2)}^{\text{osmotic pressure}} = \frac{\rho_{mixture}\, g(z_2 - z_1)}{g_c} = \frac{\rho_{mixture} RT}{mw_A}\; \ell n \frac{1}{x_A} \underset{\text{dilute}}{\cong} \frac{c_B RT}{mw_B}\overset{\text{kg solute/m}^3\text{ solution}}{} = C_B RT \tag{22–16}$$

moles solute/m³ solution

The quantity $p_1 - p_2$ is also equal to the pressure difference across the membrane and is called the *osmotic pressure*. Remember that in Eq. 22–16, A represents the permeable component, and B the nonpermeable component.

EXAMPLE 22–3.   Dissociation of ocean salt

Consider ocean water (3.48% salt) to consist only of water and salt (NaCl, $\overline{mw}$ = 58.5 gm/mol), some of which is dissociated into its positive ions and negative ions, as follows

$$NaCl \rightleftarrows Na^+ + Cl^-$$

Find the fraction of salt dissociated at 10°C at which temperature the vapor pressure of pure water is 1227.6 Pa while the vapor pressure of ocean water is

1206.5 Pa. Assume that salt and its ions form an ideal solution (questionable!) with water. Also note that the vapor pressure of salt and its ions is zero.

Solution

Relating the vapor pressure of ocean water with fresh water by the ideal solution expression of Eq. 21–2, we find for the water component of the mixture

$$p_{ocean} = p_{fresh} x_{ocean}$$

from which the mole fraction of water in ocean water is

$$x_{ocean} = \frac{p_{ocean}}{p_{fresh}} = \frac{1206.5}{1227.6} = 0.9828$$

In terms of masses and molecular weights

$$x_{ocean} = \frac{\text{moles of water}}{\text{total moles}} = \frac{\left(\dfrac{kg}{mol}\right)_{water}}{\left(\dfrac{kg}{mol}\right)_{water} + \left(\dfrac{kg}{mol}\right)_{NaCl, Na^+, Cl^-}}$$

or

$$0.9828 = \frac{0.9652 / 18}{0.9652 / 18 + 0.0348 / \overline{mw}_{salt}}$$

average for salt, Na⁺, and Cl⁻

from which the mean molecular weight of NaCl and its ions is

$$\overline{mw}_{salt} = 37.1 \text{ kg/mol}$$

Finally, let the fraction of NaCl dissociated be z. Then

$$(1 - z)\, \overline{mw}_{NaCl} + 2z\, \overline{mw}_{ions} = (1 + z)\, \overline{mw}_{avera}$$

or

$$(1 - z)\,(58.5) + 2\,(z)\left(\frac{58.5}{2}\right) = (1 + z)\,(37.1)$$

or

$$z = 0.566$$

So, on assuming an ideal solution, the fraction of NaCl decomposed into its ions is <u>56.6%</u> ←

EXAMPLE 22–4.   Osmotic pressure of sea water

Evaluate the osmotic pressure of ocean water in equilibrium with fresh water at 10°C.

Data

For ocean water:
Salinity = 3.48%, density = 1028 kg/m^3
Mean molecular weight of salt and its ions = 0.0371 kg/mol, from Example 22–3
Mole fraction of water in the mixture = 0.9828, from Example 22–3

Solution

Replacing all values known into Eq. 22–16 gives

$$\Delta p = \frac{(1028)\,(8.314)\,(283)}{0.018}\,\ell n\,\frac{1}{0.9828} = 2.33 \times 10^6 \text{ Pa}$$

$\underset{osmotic}{\curvearrowright}$
$$= \underline{23 \text{ atm}} \quad \longleftarrow$$

This is equivalent to a column of sea water Δz high, where from Eq. 22–16,

$$\Delta z = \frac{\Delta p(g_c)}{\rho_{mixture}(g)} = \frac{(2.33 \times 10^6)\,(1)}{1028\,(9.8)} = \underline{\underline{231 \text{ m}}} \quad \longleftarrow$$

Alternative solution

Using the approximation shown at the right of Eq. 22–16 we get

$$C_{solute} = \left(0.0348\ \frac{\text{kg salt}}{\text{kg solution}} \right) \left(1028\ \frac{\text{kg solution}}{\text{m}^3\ \text{solution}} \right) = 35.77\ \frac{\text{kg salt}}{\text{m}^3\ \text{solution}}$$

Therefore, from Eq. 22–16

$$\Delta p = \frac{(35.77)\,(8.314)\,(283)}{0.0371} = 2.27 \times 10^6 \text{ Pa}$$

$$= \underline{\underline{22.4 \text{ atm}}} \quad \longleftarrow$$

D. OSMOTIC WORK AND POWER

Figure 22–8 illustrates the meaning of this equilibrium across the semipermeable membrane.

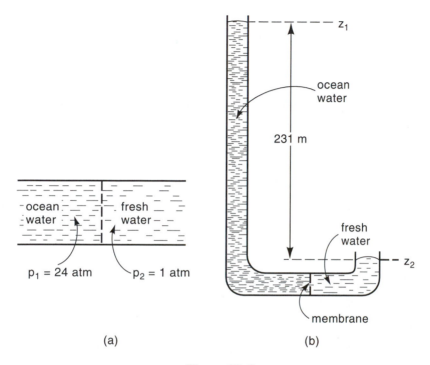

(a)   (b)

Figure 22–8.

The reversible work needed to produce 1 kg of fresh water from ocean water requires that we compress the ocean water to a bit over 24 atm (see Figure 22–8a), in which situation Eq. 13–15 becomes

$$w_{sh,1\to2} = -\int_1^2 v_{ocean}\,dp = \frac{p_1 - p_2}{\rho_{mixture}} \quad \text{[J/kg]} \tag{22–17}$$

Alternatively, this work requires that we raise 1 kg of ocean water just over 231 m (see Figure 22–8b) in which case Eq. 22–9 or 5–2 becomes

$$w_{sh,1\to2} = -\Delta E_p = -\frac{m_{mixture}g(z_2 - z_1)}{g_c} \quad \text{[J/kg]} \tag{22–18}$$

The power requirement is then found by combining Eqs. 22–16, 22–17, and 22–18 to give

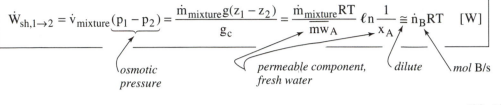

$$\dot{W}_{sh,1\rightarrow2} = \dot{v}_{mixture}\underbrace{(p_1 - p_2)}_{\substack{osmotic\\pressure}} = \frac{\dot{m}_{mixture}\,g(z_1 - z_2)}{g_c} = \frac{\dot{m}_{mixture}RT}{mw_A}\ell n\frac{1}{x_A} \cong \dot{n}_B RT \quad [W]$$

osmotic pressure   *permeable component, fresh water*   *dilute*   *mol B/s*

(22–19)

In practice imposed pressures of about 100 atm are used to produce fresh water from ocean water at a reasonable flow rate.

EXAMPLE 22–5.   Osmotic power of rivers

How much power can be extracted by reversible salination of the Hudson River at New York City?

Data

The flow of the river averages 610 m³/s; from the U.S. Geological Survey Bulletin #44, 1949.

Solution

We must first find the volume of ocean water that contains 610 m³ of fresh water. So

$$v_{mixture} = (610 \text{ m}^3 \text{ fresh})\left(\frac{1000 \text{ kg fresh}}{1 \text{ m}^3 \text{ fresh}}\right)\left(\frac{1 \text{ kg ocean}}{0.9652 \text{ kg fresh}}\right)\left(\frac{1 \text{ m}^3 \text{ ocean}}{1028 \text{ kg ocean}}\right)$$

from Example 22–4

$$= 615 \text{ m}^3 \text{ ocean}$$

Method 1   From osmotic pressure

The power produced when high pressure liquid goes reversibly to a low pressure is given by Eq. 22–19 as

$$\dot{W}_{rev} = \dot{v}_{mixture}(-\Delta p)$$

from Example 22–4

$$= \left(615 \frac{\text{m}^3}{\text{s}}\right)(23 \text{ atm})\left(\frac{101\,325 \text{ Pa}}{1 \text{ atm}}\right) = \underline{\underline{1.43 \times 10^9 \text{ W}}} \quad \longleftarrow$$

Method 2   From the developed head

Just let the fresh water drop 231 m reversibly. Then, at the membrane it will be in equilibrium with the salt water. Equation 22–19 in this case becomes

$$\dot{W} = \frac{\dot{m}_{mixture}g(-\Delta z)}{g_c}$$

$$= \frac{\left(615\ \frac{m^3}{s}\right)\left(\frac{1028\ kg}{m^3}\right)\left(9.8\ \frac{m}{s^2}\right)(231\ m)}{1\ \frac{kg\ m}{s^2 N}} = \underline{\underline{1.43 \times 10^9\ W}} \quad \longleftarrow$$

E.   THE LESSONS OF THERMO

Thermo tells that if you approach equilibrium irreversibly (inefficiently, with an increase in total entropy) you will be extracting less useful work than if you do it efficiently. As examples:

1. When heat flows from a hot source to a cold sink, the Carnot engine does it most efficiently (see Chapter 18).
2. When high pressure gas is led efficiently to a low pressure, you can get work from this operation, work with which you can even pump heat "uphill" from a lower to higher temperature.
3. When the gases A and B mix at a pressure π, work can be extracted if this is done efficiently.

   In all these cases of heat flow from high to low temperature, of gas flow from high to low pressure, or of mixing of gases, if this is done irreversibly, then you are losing available useful work.

   If we apply this concept to the mixing of fresh water and ocean water, we see that if we are clever and do it reversibly, then we can extract an enormous amount of work—in fact, the equivalent of a 231 m high waterfall or dam.

   How does sap rise 100 m up a tree? Could it not be that tree roots act as semi-permeable membranes, with water and organics inside and rainwater outside?

1.   River Meets Ocean

Let us play with this idea. Suppose we wish to extract work when fresh river water flows into the ocean. In principle, this can be done with two dams shown in Figure 22–9.

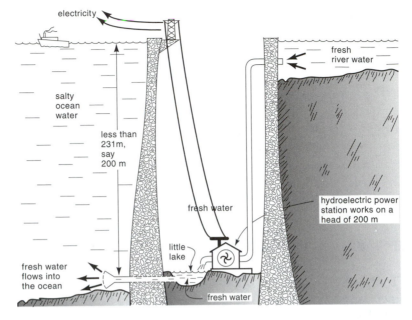

Figure 22–9.

But why use two dams when one will do (Figure 22–10)?

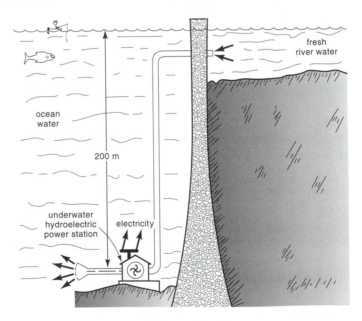

Figure 22–10.

Or better still, why not get rid of the dams altogether? Just keep the two waters separated by a plastic sheet (Figure 22–11).

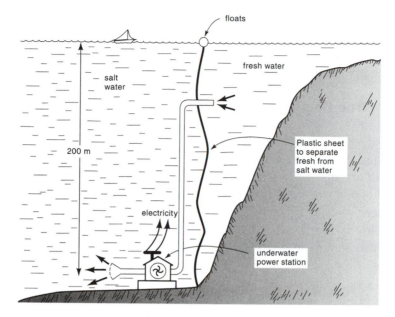

Figure 22–11.

Or even get rid of the sheet altogether (Figure 22–12).

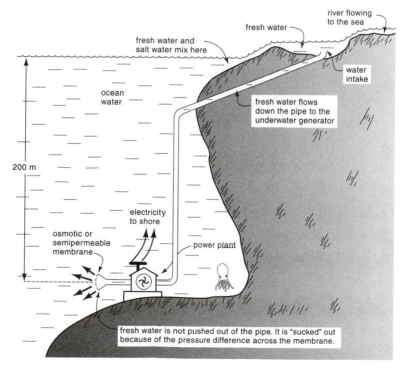

Figure 22–12.

This discussion is based on theoretical consideration. In practice all sorts of inefficiencies enter the picture—nonideal membranes, silt and solids in fresh water to plug the membranes, slow flow across the membranes—that make this concept not quite economical today.

2.   The Real Ocean

As an afterthought, the real ocean is in close to thermal equilibrium; however, it is not in chemical equilibrium. As with Example 22–2, if the surface concentration of salt is 3.5%, then it should be about 8% down deep at 10 km. But it is not—it is roughly 3.5% throughout. The reason for this is that the circulation of ocean waters (Figure 22–13), from arctic to equator and back (about 2 million years) is much faster than the diffusional process that leads to the equilibrium concentration profile (more than 100 million years).

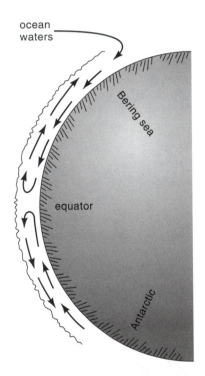

Figure 22–13.

So, if we are clever, we could in principle extract energy by helping the ocean approach equilibrium. Levenspiel and de Nevers (*Science, 183,* 157 (1974)) show how, instead of extracting work, one can use this work to separate fresh water from the ocean at no cost with a very simple device, just a long, long pipe with an ideal semipermeable membrane at its bottom (Figure 22–14).

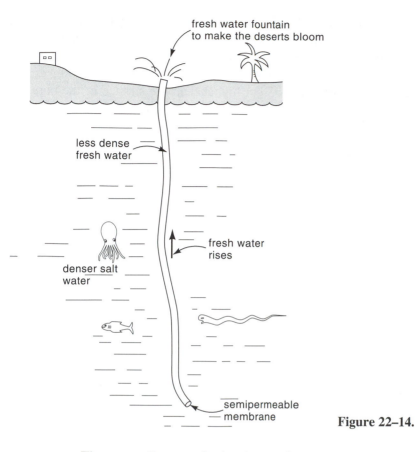

Figure 22–14.

There are all sorts of other interesting possibilities around us. Remember—when a system is not in equilibrium, a clever engineer should be able, in principle, to extract energy from the system while inviting it to approach equilibrium.

PROBLEMS

1. The measurement of osmotic pressure of solutions containing proteins and other macromolecules is the best way for determining the molecular weight of these giant molecules. Here is an example.

Four grams of hemoglobin are introduced into a measuring flask, which is then topped with water to 100 cm^3. The osmotic pressure of this solution relative to pure water is then measured and found to be 1600 Pa at 25°C. From this experiment determine the molecular weight of hemoglobin.

2. A solution of polymer in benzene is made by dissolving 1 gm of polymer in 100 ml of benzene ($\overline{mw}$ = 0.078 kg/mol, ρ = 879 kg/m^3) to make a 1 wt % solution. This solution is on one side of a U-tube, separated by a semipermeable membrane from pure benzene on the other side. At equilibrium at 25°C the surface of the solution is 3 cm higher than the surface of the pure benzene. What is the molecular weight of the polyethylene?

CHAPTER 23

CHEMICAL REACTION EQUILIBRIUM

Before we start, note the following: Ignoring E_p and E_k, Figure 23–1 shows what happens when a system moves toward equilibrium.

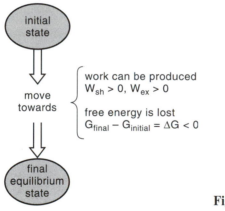

Figure 23–1.

When a system moves "uphill" away from equilibrium, work needs to be done on the system, and it gains free energy. Keep this in mind as you go through this chapter.

A. REACTION OF GASES

When chemicals react to equilibrium, you can get maybe very slight conversion to product, maybe 50%, maybe practically complete conversion. It is this conversion—and how pressure, temperature, and composition of feed affect it—that interests us. Thermo has things to say about this, so let us examine this question.

Let us start by considering the simple reaction of A to R

$$aA \rightleftarrows rR \tag{23-1}$$

taking place at equilibrium at fixed temperature and pressure in the chemical reactor shown in Figure 23–2, and let p_{Ae} and p_{Re} be the equilibrium partial pressures of the components in the vessel.

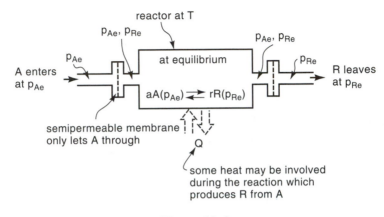

Figure 23–2.

If we add some A (at p_{Ae}) to the reactor, some R (at p_{Re}) will leave; if we add some R (again at p_{Re}), then some A (at p_{Ae}) will leave. If the reactor stays at equilibrium, the work needed to do either of these changes is zero, so the free energy change from Eq. 22–10 is

$$\Delta G_{\text{A to R, or R to A}} = 0$$

Let us now consider that at equilibrium both p_{Ae} and p_{Re} are smaller than 1 atm. (With no loss in generality we could have assumed pressures higher than 1 atm.) Suppose the feed stream for reactant A comes to the reactor at a so-called standard condition, say at $p_A^o = 1$ atm, and suppose that some R is removed at its standard condition, again at $p_R^o = 1$ atm. Then some expansion and compression work is required on the flow streams. This is shown in Figure 23–3.

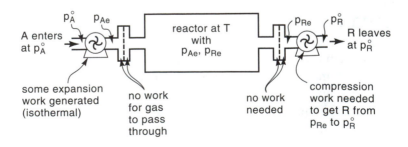

Figure 23–3.

So the shaft work produced (expansion and compression) when a moles of A are converted to r moles of R, all at temperature T (no T_0 involved), is given by

$$W_{sh,produced} = W_{ex} = -\Delta G° = aRT \ \ell n \frac{p_A^°}{p_{Ae}} + rRT \ \ell n \frac{p_{Re}}{p_R^°}$$

we call this — *the standard free energy for this reaction*

Eq. 13–16, for isothermal expansion

Eq. 13–16, for isothermal compression

$$= RT \ \ell n \frac{\left(\dfrac{p_{Re}}{p_R^°}\right)^r}{\left(\dfrac{p_{Ae}}{p_A^°}\right)^a} \qquad (23\text{–}2)$$

$$= RT \ \ell n \ K$$

More generally, for any ideal gas mixture reacting at temperature T, for example

$$aA \rightleftarrows rR + sS \ldots \Delta G° \qquad (23\text{–}3)$$

we have

standard free energy of reacting components from their elements

$$\Delta G° = rG_{f,R}^° + sG_{f,S}^° - aG_{f,A}^°$$

$$= -RT \ \ell n \ \frac{\left(\dfrac{p}{p^°}\right)_R^r \left(\dfrac{p}{p^°}\right)_S^s}{\left(\dfrac{p}{p^°}\right)_A^a} = -RT \ \ell n \ K \qquad (23\text{–}4)$$

all $p° = 1$ atm

called the equilibrium constant for reaction

where

$$K = \frac{K_p}{(p° = 1 \ atm)^{\Delta n}} = \frac{K_y \pi^{\Delta n}}{(p° = 1 \ atm)^{\Delta n}} = \frac{K_C (RT)^{\Delta n}}{(p° = 1 \ atm)^{\Delta n}}$$

$$K_p = \frac{p_R^r \ p_S^s}{p_A^a}, \quad K_y = \frac{y_R^r \ y_S^s}{y_A^a}, \quad K_c = \frac{C_R^r \ C_S^s}{C_A^a} \qquad (23\text{–}5)$$

and

$$\Delta n = r + s - a, \quad C_i = p_i RT$$

EXAMPLE 23–1.   Finding equilibrium compositions

Find the equilibrium partial pressures of all components for the following reaction taking place in a constant pressure flow reactor at 298 K

$$\left. \begin{array}{l} p_{A0} = 1.5 \text{ atm} \\[4pt] p_{B0} = 1.5 \text{ atm} \end{array} \right\} \quad A + B \rightarrow R + S \quad \Delta G^\circ{}_{298} = 0$$

Solution

For $\Delta G^\circ = 0$ Eq. 23–4 shows that

$$K = e^{-\Delta G^\circ / RT} = e^{-0/8.314(298)} = 1$$

$$= \frac{p_{Re} p_{Se}}{p_{Ae} p_{Be}} \Big/ (p^\circ)^{\Delta n} = \frac{p_{Re}\, p_{Se}}{p_{Ae} p_{Be}}$$

Now make an accounting of the moles involved, starting with, say, 3 moles; actually, any number of moles will do:

	A	B	R	S	total
initial moles	1.5	1.5	0	0	3 mol
moles at equilibrium	1.5-x	1.5-x	x	x	3 mol
mole fraction at equilibrium	$\dfrac{1.5\text{-}x}{3}$	$\dfrac{1.5\text{-}x}{3}$	$\dfrac{x}{3}$	$\dfrac{x}{3}$	1
partial pressure at equilibrium	1.5-x	1.5-x	x	x	3 atm

So replacing these values in the expression for K

$$K = \frac{p_R\, p_S}{p_A\, p_B} = \frac{(x)(x)}{(1.5 - x)(1.5 - x)} = 1$$

or

$$\frac{x}{1.5 - x} = 1 \quad \text{or} \quad x = 0.75$$

Therefore

$$\left. \begin{array}{l} p_{Ae} = 1.5 - x = 0.75 \text{ atm} \\[4pt] p_{Be} = 0.75 \text{ atm} \\[4pt] p_{Re} = 0.75 \text{ atm} \\[4pt] p_{Se} = 0.75 \text{ atm} \end{array} \right\} \quad \longleftarrow$$

ΔG° and Temperature

Chapter 22 showed that for a system at constant temperature, $\Delta E_p = \Delta E_k = 0$

$$\Delta G = \Delta H - T\Delta S$$

and since ΔH and ΔS are functions of temperature, ΔG also varies with temperature. So we usually designate the temperature we are dealing with, such as ΔG°_{298K}, ΔG°_{500K}, and so on.

Additivity of ΔG° values

Chapter 9 showed that ΔH can be summed for the various reaction components. The same is true with ΔG°. So values of ΔG°_f for the formation of chemicals from their elements are listed in Table 23–1. These values can be used to find the free energy of reaction of these materials.

TABLE 23–1   Standard Free Energy of Compounds from Their Elements

Compound	ΔG°_f J/mol, at 25°C
CO	−137 280
CO_2	−394 380
CH_4	−50 790
C_3H_6	+62 720
C_3H_8	−23 490
$C_2H_5OH(g)$	−168 620
$C_2H_5OH(\ell)$	−174 770
C_6H_6	−8 200
HCN	+120 080
$H_2O(g)$	−228 590
$H_2O(\ell)$	−237 190
$H_2S(\ell)$	−27 363
N_2O	+103 600
MgO	−569 570
$SO_2(g)$	−300 369
$SiO_2(s)$	−805 002

Composition and Conversion at Equilibrium

Here is how we determine composition and conversion at equilibrium:

1. From Table 23.1 evaluate ΔG°_f for all the reaction components.
2. Evaluate ΔG° for the reaction.

3. Calculate the equilibrium constant K from Eq. 22–4.
4. Noting the standard conditions chosen for ΔG_f^0, determine the partial pressure of the various reaction components.

The Meaning of Large Positive and Negative ΔG°

Suppose your table of free energies tells you that the independent reactions x, y, and z can go as follows:

$$\text{reaction x:} \quad A + B \rightleftarrows R + S \quad \Delta G° = -22\ 819 \text{ J}$$
$$\text{reaction y:} \quad C + D \rightleftarrows T + U \quad \Delta G° = 0$$
$$\text{reaction z:} \quad E + F \rightleftarrows V + W \quad \Delta G° = +22\ 819 \text{ J}$$

Then, for an equimolar feed mixture, and noting that $\ell n\ K = -\Delta G°/RT$, we have at equilibrium,

for reaction x: $K = 10\ 000$,   and conversion to product = 99%

for reaction y: $K = 1$,   and conversion to product = 50%

for reaction z: $K = 0.0001$,   and conversion to product = 1%

These numbers tell something useful, that at a given temperature

> a large negative $\Delta G°$ means a high equilibrium conversion
>
> a large positive $\Delta G°$ means negligible conversion

However, when conversions at different temperatures are evaluated, you must compare values of $\Delta G°/RT$ at these different temperatures, not just $\Delta G°$.

Simultaneous Reactions

When reaction between components can occur in a number of ways, then one has a more complicated algebraic problem to solve. For example, in the important industrial process for producing hydrogen, the reactants methane and steam are passed over a catalyst at about 600°C. At these conditions thermo tells that the following reactions plus many others could occur:

$$CO + H_2O = CO_2 + H_2 \quad \Delta G° = -6000 \text{ J}$$
$$CH_4 + H_2O = CO + 3H_2 \quad \Delta G° = +4000 \text{ J}$$
$$2CH_4 = C_2H_6 + H_2 \quad \Delta G° = +70\ 000 \text{ J}$$
$$H_2O = H_2 + 1/2 O_2 \quad \Delta G° = +200\ 000 \text{ J}$$
$$CO_2 = CO + 1/2 O_2 \quad \Delta G° = +206\ 000 \text{ J}$$

Considering these standard free energy values, we can safely ignore the last three reactions and only deal with the first two. However, with today's high-speed computers with their special programs you can treat any number of reactions and let the computer do all the hard work for you.

B. HETEROGENEOUS REACTIONS INVOLVING GASES, LIQUIDS, SOLIDS, AND SOLUTIONS

The treatment of heterogeneous reacting systems follows the above plan except that the standard states are chosen as follows:

> Gases—pure component at 1 atm and 25°C
> Solid—pure solid in its usual state and usual pressure at 25°C
> Liquid—pure liquid at its vapor pressure at 25°C
> Solute in liquid—at its partial pressure in a 1 molal solution at 25°C

For the heterogeneous reaction $aA + bB \rightarrow rR + sS$ the equilibrium constant can be written as

$$K = \frac{a_R^r a_S^s}{a_A^a a_B^b} \tag{23-6}$$

where a_i is called the activity of component i.

- For component i of an ideal gas, Eq. 23–4 shows that

$$a_i = \frac{p_i}{p_i^\circ} = \frac{p_i}{(1\ atm)} \tag{23-7}$$

- For pure liquids and solids

$$a_i = \frac{p_i}{p_i^\circ} \underset{=}{\overset{at\ 298K}{=}} 1 \tag{23-8}$$

- For solutes of molality M (moles per liter) in ideal solutions, Raoult's law tells that

$$a_i = \frac{p_i(at\ molality\ M)}{p_i(at\ M = 1)} = M$$

so $a_i = 2$ for a 2 mol/lit solution, $a_i = 5$ for a 5 mol/lit solution.

EXAMPLE 23–2.   Multiple heterogeneous reactions

At 527°C reactant A can decompose in two ways

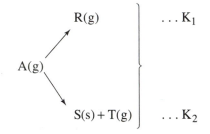

We pass a gaseous feed through a long clean stainless steel tube kept at 527°C. If equilibrium is reached,

    (a) What would be the composition of the exiting gas stream?

    (b) How much solid, if any, is deposited per mole of entering feed?

The feed consists of pure A, $p_{A0} = 1$ atm and $K_1 = 1$, $K_2 = 2$

Solution

Let us first determine whether solid deposits or not. For this, guess that it either does or doesn't and see whether the result makes sense.

Assume no deposit of solid, or $a_S < 1$. Then only the first reaction goes, or

$$K_1 = \frac{p_{Re}/p_{Ae}}{(p°)^{\Delta n}} = \frac{p_{Re}}{p_{Ae}} = 1$$

$$\Delta n = r\text{-}a = 0$$

or $p_{Ae} = 1/2$ atm $p_{Re} = 1/2$ atm

Now check with the second reaction. Note, if no solid forms, then $p_T = 0$

$$K_2 = \frac{a_S\, p_{Te}}{p_{Ae}} = \frac{a_S(0)}{1/2} = 2 \quad \dots \text{ or } a_S = \infty$$

$a_S > 1$ means that solid *will* deposit, and this contradicts our original assumption. So our original assumption is wrong.

Assume that solid does deposit, or $a_S = 1$.

    Let the fraction of A reacting in the first reaction be x and the fraction in the second reaction be y. Then make an accounting of the materials.

	A	R	S	T	total in gas
Initial moles	1	0	0	0	1 mol
Moles at equilibrium	1-x-y	x	y	y	1 mol
p_e of components	1-x-y	x		y	1 atm

For the two reactions

$$K_1 = \frac{p_{Re}}{p_{Ae}} = \frac{x}{1-x-y} = 1, \quad K_2 = \frac{a_S \, p_{Te}}{p_{Ae}} = 2$$

with $a_S = 1$

Solving simultaneously for x and y gives

$$x = {}^1/_4 \text{ and } y = {}^1/_2$$

so

$$p_{Ae} = 1 - x - y = {}^1/_4 \text{ atm}$$

$$p_{Re} = x = {}^1/_4 \text{ atm}$$

$$p_{Re} = y = {}^1/_2 \text{ atm}$$

$$\text{Solid deposit} = y = {}^1/_2 \text{ mol S/mol A fed}$$

C. LIMITATION OF THERMO AND TRICKS TO MAKE REACTIONS GO

Trick 1

Although thermo may predict, say, 60% conversion of reactant at equilibrium, the rate of reaction may be so slow that the reaction may not proceed at all. For example, if you put hydrogen and oxygen in a tank at room temperature, thermo predicts practically complete conversion to water,

$$H_2 + {}^1/_2 O_2 \longrightarrow H_2O \qquad \Delta G° = -228\,590 \text{ J}$$

but in practice you'd have to wait for many thousands of years to see any measurable conversion.

A catalyst is quite specific. It can accelerate just one reaction, not another. Thus, with the right catalyst you can get your desired reaction to go. For the above reaction, specially prepared platinum is an appropriate catalyst, and will make this reaction go to completion practically instantaneously.

Trick 2

Suppose you have a reaction A $\rightleftarrows$ R with an equilibrium conversion of just 10%. You can get a very high conversion if you can remove R from the reacting mixture. This will cause more A to react to R, so as to get back to equilibrium.

EXAMPLE 23–3. ΔG^0 of a remarkable gas-liquid reaction

Example 9–2 showed how to kill the enemy humanely. But I have a simpler and cheaper way of preparing this potent mixture. Simply bubble air through white wine spiced with a pinch of a special catalyst, and out comes the desired deadly but lovable gas. I haven't yet developed the catalyst but I am working on it, and it should be no problem. The stoichiometry of the reaction is

$$C_2H_5OH \text{ (aqueous)} + 3/2\ O_2 + 2N_2 \rightleftarrows 2HCN \text{ (g)} + N_2O \text{ (g)} + H_2O \text{ (}\ell\text{)}$$

$$\underbrace{}_{white\ wine} \quad \underbrace{}_{from\ air} \quad \underbrace{}_{product\ gas} \tag{23–9}$$

I wonder what the free energy of this reaction at 25°C is. With the right catalyst do you think that this reaction would go? Do not try to calculate the conversion.

Solution

Start by making a reaction map (Figure 23–4) showing the free energy of formation of the reaction components.

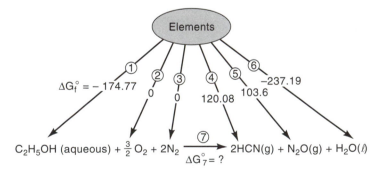

Figure 23–4.

So for the reaction of Eq. 23–9

$$\Delta G^{\circ}_{f1} + \Delta G^{\circ}_{f2} + \Delta G^{\circ}_{f3} + \Delta G^{\circ}_{7} = \Delta G^{\circ}_{f4} + \Delta G^{\circ}_{f5} + \Delta G^{\circ}_{f6}$$

Replacing values in kJ/mol,

$$\Delta G°_7 = 2\,(120.08) + 103.6 - 237.19 - 2.06\,(-174.77 - 0 - 0)$$

$$= 466.60\ \text{kJ}$$

see note below

So we find

$$\Delta G°_7 = +460\ 600\ \text{J}$$

$$K = e^{-\Delta G°/RT} = e^{-460\ 600/8.314(298)} = 1.8 \times 10^{-81}$$

With so small an equilibrium constant this reaction is impractical. Try some other method.

Note. As illustration, let us evaluate the activity of the wine. Assume a 12 volume % of alcohol in water, and ignore all other ingredients. With an alcohol density of 789 kg/m³ we have

$$\text{For the wine:}\quad (0.120\ \text{vol. fraction})\left(789\,\frac{\text{gm}}{\text{lit}}\right) = 94.6\,\frac{\text{gm alcohol}}{\text{lit}}$$

$$\text{Molality:}\quad \frac{94.6\ \text{gm/lit}}{46\ \text{gm/lit}} = 2.06\,\frac{\text{mol}}{\text{lit}}$$

So the activity of the wine is

$$a_{\text{wine}} = 2.06$$

This value is used in the solution above.

D. LIVING CREATURES AND THERMO

We may feel quite pleased with ourselves because in the last 200 years we have learned how to extract work when heat flows from a hot source to a cold sink (Figure 23–5).

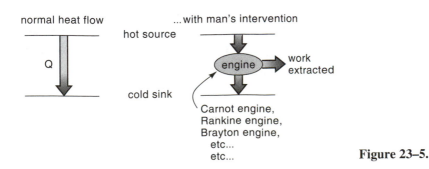

Figure 23–5.

But these efforts are puny compared with nature's accomplishments in this line. Its creatures today extract useful work and life energy by coaxing materials with large free energy to react and produce product having small free energy. For example, man and animals extract their life energy by digesting food (high G) to form products CO_2 and H_2O (low G) (Figure 23–6).

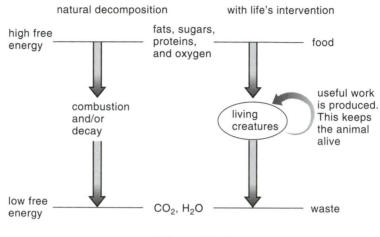

Figure 23–6.

In fact, whatever the environment, if a high free energy can be run down, then life can establish itself and thrive. For example, in earth's early days the atmosphere was rich in carbon dioxide and hydrogen, with no oxygen, so life forms developed that fed on these materials.

Today, deep in the ocean and far from sunlight we find hot hydrothermal vents that release waters rich in sulfur compounds that are poisonous to us. However, these salts have a high free energy so these vents support colonies of microbes, each living in its favored environment, some at scalding hot temperatures, others at progressively lower temperatures.

Whether in the hot hydrothermal vents in the ocean deeps, or in the boiling mineral rich volcanic waters of Yellowstone National Park—in fact, anywhere where the material has a high free energy—life forms will adapt and move in. However, this will only occur if the food has a higher free energy than does the product. Recall the relation of useful work and free energy from Chapter 22.

$$W_{sh} = -(\Delta G^\circ + \Delta E_p + \Delta E_k)$$

work available   drop in free energy

(22–4)

EXAMPLE 23–4. The world of science fiction

Science fiction writers have visualized worlds in which "creatures" lived on silicon compounds in place of carbon compounds. They ate sand and excreted pure silicon and oxygen.

$$SiO_2 \text{ (s)} \longrightarrow Si(s) + O_2(g)$$

Does this make sense from the point of view of thermo?

Solution

Let us see whether it is possible to extract useful work from this reaction. For this, make the sketch shown in Figure 23–7.

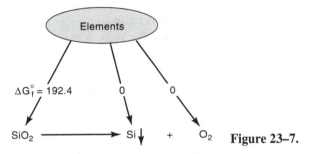

Figure 23–7.

So for the reaction

$$\Delta G° = 0 + 0 - (-192\ 400) = 192\ 400 \text{ J/mol}$$

It is not possible to obtain any useful work from this reaction because the final free energy is greater than the initial, or

$$-\Delta G° = +W_{sh} < 0 \longleftarrow$$

so it is not possible for this life form to exist.

PROBLEMS

1. I wish to pump A from here to there at 25°C and 1 atm. However, I am afraid that A may decompose as follows

$$A(g) \rightleftarrows R(g) + H_2$$

What fraction could be expected to decompose? Are my fears justified?

Data   The standard free energies of formation are

For gaseous A at 25°C: $G°_{A,298K} = 10\ 000 \text{ J}$

For gaseous R at 25°C: $G°_{R,298K} = 40\ 000 \text{ J}$

2. This is a message I received on my telephone answering machine:
"Hi, Octopus, I feel pretty good today, and here's why. Last night at Goofy's Tavern a scientist type casually let drop that he had cooked up a white powder that catalyzed the decomposition of propane gas to propylene and hydrogen

$$C_3H_8(g) \rightleftarrows C_3H_6(g) + H_2$$

The fellow never realized what this meant. But I, with my natural business sense, immediately grasped its significance. You see, propylene is the raw material for polypropylene plastic and is four times as expensive as propane, a mere fuel. Imagine, I will get on the phone:
    'Hello, Northwest Propane? Send me $100 000 worth of propane.'
I pass it across the powder and presto . . . another phone call:
    'Hello, DuPont? How would you like $400 000 worth of propylene?'
And a flaskful of this magic powder only cost me $1000. What a killing, ho! ho! Now pay attention. Since I am short of cash, and since you are my good friend, I'll do you a favor. I'll let you in on 50% of the action for a mere $100 000. Just write out a check for this amount and mail it to me tomorrow."

—-(end of phone message)—-

Question Before writing out the check how about evaluating the proposed process from the thermodynamic point of view. What percent conversion of propane to propylene could possibly be obtained? What do you conclude?

3. Reactant A decomposes according to the stoichiometry

$$A(g) = R(g) + S(g)$$

In a flow reactor fed pure A at 1 atm and 298°K, A decomposes to form an equilibrium mixture of A, R, and S, in which $p_{Ae} = 0.25$ atm.

(a) What percent decomposition of A does this correspond to?

(b) Find the equilibrium constant for this reaction.

(c) Find $\Delta G°$ for the reaction $2R + 2S \rightleftarrows 2A$.

(d) Find the partial pressure of A in the equilibrium mixture leaving a reactor if the total pressure is kept at $\pi = 10$ atm (instead of 1 atm).

Find the equilibrium partial pressure of all components for the following reactions taking place in a constant pressure flow reactor, given the following conditions:

4. ... $\left.\begin{array}{l} p_{A0} = 1 \text{ atm} \\ p_{B0} = 9 \text{ atm} \end{array}\right\}$ $A + B \rightarrow R + S \ldots \Delta G^\circ = 0$

5. ... $\left.\begin{array}{l} p_{A0} = 1 \text{ atm} \\ p_{B0} = 1 \text{ atm} \end{array}\right\}$ $A + B \begin{array}{l} \nearrow R + S \ldots \Delta G^\circ = 0 \\ \searrow T + U \ldots \Delta G^\circ = 0 \end{array}$

6. ... $\left.\begin{array}{l} p_{A0} = 2 \text{ atm} \\ p_{B0} = 2 \text{ atm} \end{array}\right\}$ $A + B \begin{array}{l} \nearrow R + S \ldots \Delta G^\circ = 0 \\ \searrow T + U \ldots \Delta G^\circ = -5443.8 \text{ J} \end{array}$

7. ... $\left.\begin{array}{l} p_{A0} = 1 \text{ atm} \\ p_{B0} = 1 \text{ atm} \end{array}\right\}$ $A + B \begin{array}{l} \nearrow R + S \ldots \Delta G^\circ = 0 \\ \searrow T + U \ldots \Delta G^\circ = 34\,225 \text{ J} \end{array}$

8. ... $p_{A0} = 1 \text{ atm}$ $2A \rightarrow R$ $\Delta G^\circ = 0$

---

9. At 161°C gaseous A can decompose as follows

$$A(g) \rightarrow B(g) + S(s) \ldots \Delta G = -2500 \text{ J}$$

Per mol of A introduced into a flow reactor at 161°C and 1 atm, and for reaction going to equilibrium, how much solid, if any, is formed?

10. At 409 K the following equilibrium exists

$$A(g) = R(g) + S(s) \qquad \Delta G^\circ{}_{409K} = 2357 \text{ J}$$

A mixture of 50% A – 50% R (mol basis) enters the reactor at 2 atm and 409 K. What is the composition of the exit stream assuming equilibrium there?

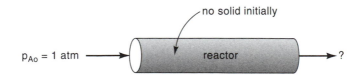

no solid initially

$p_{Ao} = 1 \text{ atm} \longrightarrow$ reactor $\longrightarrow$?

11. Solve Example 2 for a feed of pure A at 1 atm with $K_1 = 1$, $K_2 = 1$.

12. Solve Example 2 for a feed consisting of $p_{A0} = 1 \text{ atm}$ and $p_{T0} = 1 \text{ atm}$, with $K_1 = 1$ and $K_2 = 1$.

13. At given temperature and 1 atmosphere pressure we wish the reaction

$$A(g) = R(g) \ldots K_1 = 3$$

to proceed to equilibrium. Unfortunately, the side reaction

$$A(g) = S(s) + T(g) \ldots K_2 = 1$$

can also occur. We wish to avoid formation of S, and maybe adding T to the feed may help. With this in mind find what should be the feed to the reactor so that no solid will form. Also find the exit equilibrium composition of gases leaving the reactor at this condition.

At 527°C reactant A can decompose in two ways

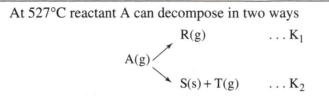

$$A(g) \nearrow^{R(g) \qquad \ldots K_1}_{\searrow S(s)+T(g) \qquad \ldots K_2}$$

We pass a gaseous feed through a fluidized bed of solid S at 527°C. If equilibrium is reached

(a) What would be the composition of the exiting gas stream?

(b) How much solid is deposited or used up per mole of entering gas feed?

Whatever happens note that solid S is always present.

14. . . . Feed consists of pure A at 1 atm, and $K_1 = 3$, $K_2 = 1$. Also $\pi \cong 1$ atm through the reactor.

15. . . . Feed consists of 50% A − 50% T (mole basis of course), $\pi \cong 2$ atm, $K_1 = 1$, $K_2 = 1$.

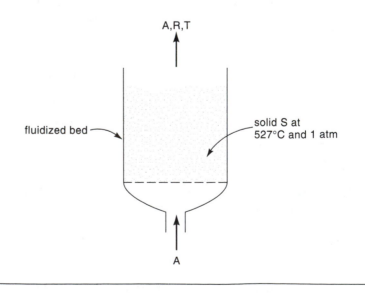

---

16. I pass pure oxygen at 1 atmosphere through a bed of carbon kept at 700°C. The reactions I think could occur are

$$C(s) + O_2 \quad = CO_2 \quad K_1$$
$$C(s) + 1/2 O_2 = CO \quad K_2 = 3 \times 10^{10}$$
$$CO + 1/2 O_2 = CO_2 \quad K_3 = 1.2 \times 10^9$$
$$C(s) + CO_2 \quad = 2CO \quad K_4 = 2$$

(a) From the above data calculate K_1.

(b) What would be the composition of the exit stream at equilibrium?

17. In the previous problem make one change—pass air (21% O_2), not pure oxygen, through the bed of red hot carbon.

18. Colored pencils are okay, but the black lead pencil is deadly, so its production and sale should be banned. Why? Because you breathe on the pencil when you write, your breath has 13% carbon dioxide, and the pencil consists of carbon. These materials react to produce CO, which we all know is deadly

$$CO_2(g) + C(s) \longrightarrow 2CO(g)$$

Students, teachers, and textbook writers are especially vulnerable. As proof of my assertion census figures show

- In 1860 (before the lead pencil was invented) the death rate in the United States was just 70 000/yr
- In 1900 it was 200 000/yr
- In 1940 it was over 1000 000/yr

You see, more pencils, more deaths. If you don't believe me, why don't you calculate the equilibrium conversion of CO_2 to CO in the air as you write?

19. Today's methane bacteria live off the energy of the reaction of carbon dioxide and hydrogen

$$CO_2 + H_2 \longrightarrow CH_4 + 2H_2O$$

Is this consistent with thermodynamics?

20. It has been suggested by some that early in Earth's life the atmosphere was rich in CO_2 and the life forms then in existence ate sulfur, breathed in CO_2, and excreted sulfur dioxide and carbon monoxide, or

$$S(s) + 2CO_2(g) \longrightarrow SO_2(g) + 2CO(g)$$

What would thermo say about this proposal?

CHAPTER 24

ENTROPY AND INFORMATION

Consider the following two statements:

A. I flipped a penny and up came a "head."

B. I'm a mushroom expert and I should tell you that the mushrooms you're now greedily devouring are poisonous, and that you'll die an agonizing death in the next hour.

I think you will agree that statement B has more "surprizal" than statement A. In other words, that the amount of information imparted by statement A is small, but is large in statement B.

Only in the twentieth century did thinkers wrestle with the ideas of "lots of information," and "little information" and try to systematically quantify the amount of information in a message. In 1948 the publication of *The Mathematical Theory of Communication* by Claude Shannon solidly linked the concept of information to the basic ideas in physics and to the concept of entropy.

Let us briefly introduce the basic concepts of information theory by considering in turn:

1. The information in a single message (you failed your last thermo test).
2. The average information in a series of messages (you got an "F," and I saw a stranger driving off in your car).
3. Total information in a series of independent messages (you got an "F," someone just stole your car, and yes, we have no bananas today).
4. The relationship between information and entropy.

A. INFORMATION IN A SINGLE MESSAGE ABOUT A SINGLE EVENT

Suppose a message is sent that says event i has occurred. How much information is given in that report? We define two terms: first, let the probability of event i actually happening be $p_{\text{event } i}$. Second, the message sent may be correct, or it may be garbled and may be incorrect. Let the probability of the message being correct be $p_{correct}$. Then the amount of information received is defined as[1]

$$
\begin{aligned}
\Delta I &= \log \frac{p_{correct}}{p_{event\ i}} \\
&= \log p_{correct} - \log p_{event\ i}
\end{aligned}
\tag{24-1}
$$

If there is no reporting error, then $p_{correct} = 1$, and this is called a noiseless message. In this situation the information received by the message is

$$
\Delta I = -\log p_{event\ i}
\tag{24-2}
$$

In these expressions any logarithm base can be chosen; however, base 2 and base 10 are found to be most useful. The amount of information in these bases is measured in units called *bits* or *decits*, respectively. As examples

$$
\Delta I = \log_2 8 = 3 \text{ bits}
$$

$$
\Delta I = \log_{10} 100 = 2 \text{ decits}
$$

To convert a logarithm from base b to base a note that

$$
\log_a x = (\log_a b)(\log_b x)
$$

thus

$$
\begin{aligned}
\log_{10} x &= 0.3010 \log_2 x = 0.4343\ \ell n\ x \\
\log_2 x &= 3.3223 \log_{10} x = 1.4428\ \ell n\ x \\
\ell n\ x &= 0.6931 \log_2 x = 2.303 \log_{10} x
\end{aligned}
\tag{24-3}
$$

and

Here are some examples to show that this definition corresponds to our intuitive notions of information

---

EXAMPLE 24–1.   I flip a coin

My eyesight is good and I am never wrong. How much information do I report when I look at the coin and shout "heads"? From Eq. 24–1, using base 2

---

[1]In 1940 Alan Turing first introduced the logarithm of the ratio of these two probabilities as the "weight of evidence."

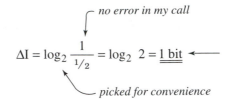

We could have chosen base 10, in which case the amount of information received is

$$\Delta I = \log_{10} \frac{1}{1/2} = \log_{10} 2 = \underline{0.3010 \text{ decit}} \longleftarrow$$

EXAMPLE 24–2.   You flipped a coin

Unfortunately, you left your glasses at home, you can't see too well, and your call is wrong on the average one time every eight calls. How much information do you give when you shout "heads"?

$$\Delta I = \log_{10} \frac{7/8}{1/2} = \log_{10} \frac{7}{4} = \underline{0.243 \text{ decits}} \longleftarrow$$

$$= \underline{0.81 \text{ bits}} \longleftarrow$$

EXAMPLE 24–3.   Throw of two dice

I correctly report that I rolled a "9" with two dice. How much information does this represent?

The number "9" can appear 6-3, 5-4, 4-5, 3-6, or 4 ways from all 36 combinations of pips, so from Eq. 24–2

$$\Delta I = -\log_{10} \frac{4}{36} = \log_{10} 9 = \underline{0.954 \text{ decits}} \longleftarrow$$

$$= \underline{3.17 \text{ bits}} \longleftarrow$$

B.   AVERAGE INFORMATION PER CALL IN A LONG MESSAGE

In a long string of reports—a, z, b, r, c, c, . . .—let symbol "a" occur with probability p_a, symbol "b" occur with probability p_b, and so on. The average information per call or symbol received, if the message is noiseless, is then

$$\overline{\Delta I} = \log \frac{p_{correct}}{p_{per\ symbol}} \overset{1}{\nearrow}$$

proper average

$$= 0 - \left(p_a \log p_a + p_b \log p_b + \ldots \right)$$

average p/symbol

$$\boxed{\overline{\Delta I}_{noiseless} = - \sum_{\substack{all\ possible \\ symbols}} p_i \log p_i}$$

(24–4)

average information per symbol received

EXAMPLE 24–4.   Many tosses of a coin

I toss a coin many times and record the results—for example, H, T, T, H, T, H, H, H, . . . How much information is given per toss? From Eq. 24–4

$$\overline{\Delta I} = -\left(p_H \log_2 p_H + p_T \log_2 p_T \right)$$

$$= -\left(\tfrac{1}{2} \log_2 \tfrac{1}{2} + \tfrac{1}{2} \log_2 \tfrac{1}{2} \right) = \tfrac{1}{2} + \tfrac{1}{2} = \underline{1\ bit} \leftarrow$$

EXAMPLE 24–5.   The roll of two dice

I roll two dice many times and record the sum of pips that show each time—for example, 3, 4, 7, 5, 9, 2, 10, ... How much information do I get, on the average, per roll? From Eq. 24–4

for numbers 2 and 12   *probability of getting a 2 or a 12*

$$\overline{\Delta I} = -\left(\frac{2}{36} \log_{10} \frac{1}{36} + \frac{4}{36} \log_{10} \frac{2}{36} + \frac{6}{36} \log_{10} \frac{3}{36} + \frac{8}{36} \log_{10} \frac{4}{36} + \frac{10}{36} \log_{10} \frac{5}{36} + \frac{6}{36} \log_{10} \frac{6}{36} \right)$$

$$= 0.0865 + 0.1395 + 0.1799 + 0.2121 + 0.2381 + 0.1297$$

$$= \underline{0.9858\ decits} \leftarrow$$

$$= \underline{3.275\ bits} \leftarrow$$

C.   TOTAL INFORMATION IN A NUMBER OF MESSAGES ABOUT INDEPENDENT EVENTS

Information about independent events is additive; hence, we can write

$$\Delta I_{total} = \sum_{\substack{\text{all independent} \\ \text{events}}} (\Delta I)_{\text{each event}} \tag{24-5}$$

EXAMPLE 24–6.   Locate a piece on a chess board

I locate a piece on a chess board by telling that it is on column B, and row 7. How much information does this represent? From Eq. 24–5

$$\Delta I_{\text{to locate}} = (\Delta I)_{\text{column}} + (\Delta I)_{\text{row}}$$

$$= \log_2 \frac{1}{1/8} + \log_2 \frac{1}{1/8}$$

$$= 3 + 3 = \underline{6 \text{ bits}} \leftarrow$$

EXAMPLE 24–7.   Molecules in a box

I have 10 gas molecules in a box, either on the left half or right half. What information do I give when I correctly say, "they are all on the left"? Let us assume that all the molecules move independently, and this is reasonable. Then Eq. 24–5 gives

$$\Delta I = (\Delta I)_{\text{molecule 1}} + \Delta I_2 + \ldots + \Delta I_{10}$$

$$= \log_2 \frac{1}{1/2} + \log_2 \frac{1}{1/2} + \ldots + \log_2 \frac{1}{1/2}$$

$$= 10 \log_2 2 = \underline{10 \text{ bits}} \leftarrow$$

D.   RELATE ENTROPY WITH INFORMATION[2]

Consider a vessel that contains $\mathbf{A} = 6.023 \times 10^{23}$ molecules of ideal gas, where $\mathbf{A}$ is Avogadro's number, the number of molecules in one mole of material. Let us assume that originally I have no idea where the molecules are. Then I correctly

---

[2]In a short technical paper published in 1929, Leo Szilard was the first to point out and show that "information" and "entropy" were two faces of the same concept.

tell you that all the molecules are in the left half of the vessel (Figure 24–1). How much information does this message represent?

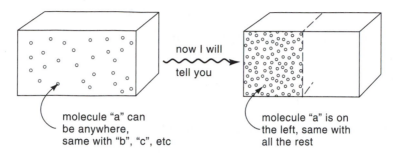

Figure 24–1.

Calculate ΔI

$$\Delta I_{total} = \sum_{\substack{all \\ molecules}} (\Delta I)_{each} = \sum_{A} \log_2 \frac{1}{1/2} = A \log_2 2 \qquad (24\text{–}6)$$

Calculate ΔS

We can also evaluate the entropy change for this process by considering an isothermal compression (Figure 24–2).

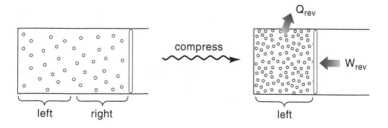

Figure 24–2.

From Eqs. 16–4 and 24–3 we write

$$\Delta s = \int \frac{dq_{rev}}{T} = \int \frac{dw_{rev}}{T} = \int \frac{pdv}{T} = \int \frac{Rdv}{v} = R \ln \frac{v_2}{v_1} \qquad (24\text{–}7)$$

$$= R \ln \frac{1/2}{1} = -R \ln 2 = -R(0.693) \log_2 2$$

Compare ΔI with Δs for the same event. From Eqs. 24–6 and 24–7 we find

$$-\frac{\Delta s}{R} = \frac{0.693 \, \Delta I}{A}$$

Now R is the gas constant per mole of ideal gas. The equivalent constant per molecule, not per mole, is called the Boltzman constant **k** where R = **k A**. Thus

$$\mathbf{k} = \frac{R}{A} = \frac{8.314 \text{ J/mol·K}}{6.023 \times 10^{23} \text{ molecules/mol}} = 1.38 \times 10^{-23} \frac{J}{\text{molecule·K}}$$

So for molecules we relate Δs with ΔI as follows

$$-\Delta s = (0.693 \, \mathbf{k})\Delta I = \frac{0.693 \, R}{A} \Delta I$$

or

$$\Delta I = \frac{A}{0.693R} \Delta s = \frac{\Delta s}{0.693k}$$
(24–8)

or

$$1 \text{ bit} = 9.56 \times 10^{-24} \frac{J}{\text{mol·K}}$$

This equation shows that the more you know about the system's molecules (where the molecules are, how fast they are going, and in what direction) the more its entropy decreases. This introduces a curious aspect to entropy.

The mass of an object is its mass, whether you've measured it or not; its temperature is at some value, whether you know it or not. But entropy is different—its value has something to do with what you know about the system. Bridgman, in *The Nature of Thermodynamics* (Cambridge University Press, 1941), put it nicely when he wrote

> It must be admitted, I think, that the laws of thermodynamics have a different feel from most of the other laws of the physicist. There is something more palpably verbal about them—they smell more of their human origin.

This aspect of thermo can and has led to much philosophizing.

EXAMPLE 24–8. An ideal gas

On the average, how long must we wait until all the molecules of one mole of gas find themselves in the left half of a container?

Data. The molecules rearrange, collide, and rearrange themselves 10^{10} times/s.

Solution

Now recall that when I know a molecule is on the left side the information gained is ΔI/molecule = 1 bit. Therefore, for one mole of gas, or A molecules, the total information gained is $\Delta I_{total} = A$ bits = 6.023×10^{23} bits. So the time that I have to wait is, on the average,

$$t = \left(\frac{6.023 \times 10^{23} \text{ changes}}{10^{10} \text{ changes/s}} \right) \left(\frac{\text{hr}}{3600\text{s}} \right) \left(\frac{\text{day}}{24 \text{ hr}} \right) \left(\frac{1 \text{ yr}}{365 \text{ day}} \right)$$

$$= 1.9 \times 10^6 \text{ years} \quad \longleftarrow$$

E. SUMMARY

In general consider the change sketched in Figure 24–3.

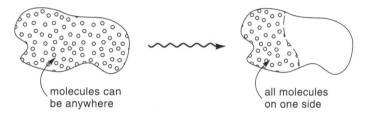

molecules can be anywhere   all molecules on one side

Figure 24–3.

This represents a change

- to a more ordered state
- away from equilibrium
- to a less probable state
- that leads to an increase in information

Thus, a movement towards equilibrium represents

- a move to a more probable state
- an increase in entropy
- a loss of information about the system
- a more disordered state

With this in mind the following quotes make sense.

> Gain in entropy always means loss of information, and nothing more. It is a subjective concept, but we can express it in its least subjective form (from G. N. Lewis, *Science, 71,* 567 (June 1930)).
>
> The movement to equilibrium is a movement from a more easily to a less easily distinguished state. (from J. Jeans)
>
> Entropy measures the missing information about a system, or our ignorance about a system's exact state.

The ideas sketched here set the stage for the development of the subjects called statistical mechanics, statistical thermodynamics, quantum thermodynamics, and quantum-statistical mechanics, with its Maxwell-Boltzman statistics, Bose-Einstein statistics, and Fermi-Dirac statistics. These statistical theories explain the gross observed behavior of the world about us in terms of the microscopic behavior in which such systems have their origins.

The great value of classical thermodynamics is due directly to the fact that the instruments we use to measure the thermodynamic variables of pressure and temperature are unable to detect the fluctuations caused by the molecular action. So, luckily for classical thermo, we are of the right size, halfway between the atoms and the stars.

EXAMPLE 24–9.   Shuffle a deck of cards

I take an ordered deck of cards (spade A, 2, 3, . . .) and I thoroughly shuffle it. How much information (in bits) have I lost, and what is ΔS for this process?

Note.   For large numbers $\ln N! \cong N \cdot \ln (N - 1)$

Solution

When I tell you that the spade A is on top of the deck, not in any of the other positions, then Eq. 24–1 tells that

$$\Delta I_{\text{spade A}} = -\log \frac{1}{1/52} = \log 52, \text{ in any units}$$

With the first card fixed, there are 51 possible places for the spade 2. When I tell you that it is position 2, then

$$\Delta I_{\text{spade 2}} = \log 51$$

and so on. So for the whole deck

$$\Delta I_{total} = \log 52 + \log 51 + \ldots + \log 1$$
$$= \log 52!, \text{ any units}$$

But when I shuffle this deck I have lost all this information, or

$$\Delta I_{shuffle} = -\log 52!$$

to convert $\log_e$
to base $\log_2$, Eq. 24–3

$$= -52[\ell n\,(52-1)](1.443)$$

$$= \underline{-221.5 \text{ bits}} \longleftarrow \text{(information lost)}$$

and

$$\Delta S_{shuffle} = +\underline{221.5 \text{ bits}} \longleftarrow \text{(more disordered)}$$

EXAMPLE 24–10.   Another shuffle

If I didn't know that the cards in the deck of Example 24–9 were originally ordered, if I shuffled the deck, then what was ΔI and ΔS for this shuffle?

Solution

Since no information was lost or gained, $\Delta S = \Delta I = 0$

Note.   Examples 24–9 and 24–10 show that ΔS doesn't just deal with a physical change. It also concerns what you know about the system. Most puzzling.

PROBLEMS

1. I flip a coin and you report "heads." However, on the average you call it incorrectly every second time. How much information do you give when you report "heads"?

2. I simultaneously flip a penny and a nickel and as in problem 1, I call "penny-heads, nickel-tails." How much information do I give per call?

3. I simultaneously flip two pennies and I call either H-H, H-T, or T-T. How much information do I give per call?

4. I start with a well-shuffled deck of 52 cards. I first write down the order of the cards on a piece of paper, then I crumple the paper, lightly salt it, and eat it. What is the entropy change for this two-step process?

5. My bicycle lock has three rings, each with six numbers—1,2,3,4,5,6. I can't understand what happened to me, but in all this excitement my mind

blanked out and I can't remember the three-number combination—1,2,6 2,4,1 6,5,6—I give up. I don't even know how much information I've lost by forgetting the combination. Could you work it out for me?

6. A, B, C, and D are four people.

   - A throws two dice, one red, the other green.
   - B is playing craps, looks at the result and says "nine."
   - C is playing backgammon and says "a six and a three."
   - D is a statistician who wants to record the frequency of dice throws, and says "a six on the red, a three on the green."

   (a) How much more information did C give than did B?

   (b) How much more information did D give than did B?

7. Two dice, one red the other green, are thrown. Alfie reports "twelve," Beetle reports "six on red and six on green." How much more information did Beetle give than Alfie?

8. Two pretty yellow canaries are confined to a quarter of their bird cage by a wire screen. The screen is removed and the canaries are free to fly anywhere in their cage. Find the entropy change that results from the removal of the screen.

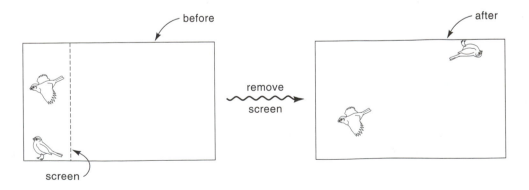

9. Three pretty canaries are happily flying around in their cage. I wait until two are on the left quarter of the cage. Then I quickly slip a screen in place catching them there. The third canary is on the right side. Find the entropy change for this process. Use any units you wish.

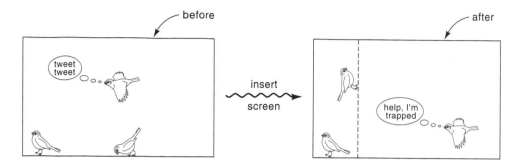

10. In a secret message the twenty-six letters of the English language are used at equal frequencies. How much information can be reported with each letter or symbol sent?

11. The letters of the alphabet in their usual frequency as used in the English language are as follows:

E = 131/1000	S = 61	U = 25	V = 8
T = 105	H = 53	G = 20	K = 3
A = 86	D = 38	Y = 20	X = 1
O = 80	L = 34	P = 20	J = 1
N = 71	F = 29	W = 15	Q = 1
R = 68	C = 28	B = 13	Z = 1
I = 63	M = 25		

The information given per letter taken singly in English frequencies is found from Eq. 24–5 to be $\Delta I = 4.70$ bits.

Now to our problem. President Jimmy Carter's Secretary of State spelled his name just the way you'd pronounce it, or

Zbigniew Szczypa Brzezinski

(a) Comparing frequencies, would you say that this is likely to be a name in the English language?

(b) Can you estimate how surprising or unusual this name is in the English language? As an example, is it as unusual as flipping fifteen heads in a row, or how many?

12. Over one hundred years ago Pasteur noticed that some chemical compounds came in two forms, left-handed and right-handed (like gloves). Not too long ago it was discovered that certain molecules in crystalline substances could be made to orient in one position or in another position—a sort of "flip" or "flop." Sometime in the future this behavior could become the basis for a new kind of information storage device.

   (a) I wonder how much information could be stored in a 1 mm cube of ice kept at $-0.1°C$ if each molecule of H_2O in the cube could flip or flop.

   (b) How does this compare to the memory of the most powerful of personal computers today, or 80 megabytes? Note that 1 byte = 8 bits of memory.

   (c) Disaster—with the memory cube in full use, the electrical power to the device is disrupted, and the cube of ice melts to water at $0°C$. What is the entropy change for this process?

13. Given N coins, all identical except for one which is heavier. With a beam balance (L-side down, R-side down, or balanced) how many weighings, n, are needed to identify the heavier coin?

14. Our hero, Bret Maverick, is doomed. He will lose this last game, his fortune, his girl, and his reputation. Dirty Dog Durango is dealing, has palmed the ace of spades to the top of the deck, and is just going to deal it to his partner in crime, Low Down Louie. Just then Bret slams down his cards, demands a cut, and takes it before Dirty Dog can even protest! Dirty Dog loses the ace, his cool, and the game. What did the cut do to the entropy of the deck as Dirty Dog knew it?

   Note.   At that time there were only 33 cards left in the deck.

15. The bird cage shown below encloses two pretty singing canaries and has slots A and B into which a cage divider can be inserted and shifted.

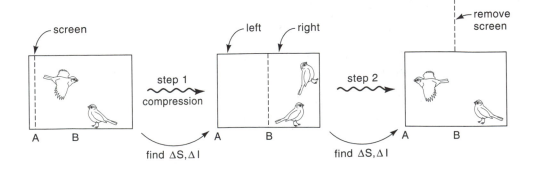

Step 1   The divider is slipped in at A and moved to the right (pushing or keeping the birds ahead of it) to position B.

Step 2   The divider is removed at B.

(a) After Step 1 what has happened to the entropy of the system and our information about the system?

(b) After Step 2 what happens to these two quantities?

Note.   This problem is a complicated version of the example treated by L. Szilard in his paper "On Entropy Reduction in a Thermodynamic System Through Intervention of an Intelligent Being" in *Zeit. fur Physik, 53* 840 (1929) where he shows that the relationship between ΔS and ΔI is not just mathematical symbolism.

CHAPTER 25
MEASURING TEMPERATURE— PAST, PRESENT, AND . . .

The origin of the thermometer, a device with some sort of scale for measuring the hotness or coldness of objects, is obscure. However, the climate in Europe in the early 1600s was ripe for it. It had to be invented at that time, and so it was . . . but by whom? In Italy Galileo had his champions, and so had Santorio, Professor of Medicine at Padua. Then there was the Welsh doctor Fludd; also the gadgeteer, inventor, and perpetual-motion-machine-maker Drebbel of Holland. But who was first? Since people in those days didn't much care about getting into the Guinness Book of Records we probably will never know. It seems that this sort of immortality was not the passion then that it is today.

In any case, by the middle-1600s the thermometer was widely known in Europe, each maker with his own scale of measurement. A popular design started with "1" in the middle of the device to represent everyday comfort. It then indicated 8 degrees of coldness and 8 degrees of hotness, each degree in turn divided into as many as 60 minutes. Isaac Newton, in an amendment to his famous "Law of Cooling," proposed a temperature scale in which the freezing point of water was taken as zero, and the temperature of the human body as 12°. Other makers were more descriptive, [viz]

Extream Hott
Very Hott
Hott
Warm
Temperate
Cold
Frost
Hard Frost
Great Frost
Extream Frost

Even as late as the mid-1800s one could buy a thermometer having eighteen different marked scales.

The development of a standardized temperature scale was a long and bumbling process. About 100 years after the invention of the thermometer, enter Daniel Gabriel Fahrenheit, Danzig born, Dutch adopted, master instrument maker, and traveller. While in Copenhagen around 1704 he visited Ole Rømer, Danish astronomer, where he observed him busily calibrating thermometers. Struck by the elegant simplicity of Rømer's choice of calibration points—7 1/2° for the ice-water and 22 1/2° for body temperature—Gabe immediately adopted these for his own. But since fractions always bothered him, he eventually multiplied everything by four to get rid of the two halves, fudged upward a bit, and ended up with 32° and 96° for these calibration points. On this scale a mixture of sea salt and ice melted at about 0° and a person's temperature was about 100°. Because humans are somewhat unreliable, some hot blooded, others questionable, before long the freezing point (32°) and the boiling point of water (212°) became the accepted calibration points. Since Fahrenheit's thermometers sold well, this scale soon became widely adopted.

While the Fahrenheit scale came into wide use in England and Holland, the French completely ignored this northern development. Their man was Réaumur, who after 1731 vigorously championed a spirits of wine thermometer—French wine, of course, and red for easy reading. His scale went from 0° for ice water to 80° for boiling water. Unfortunately, however, among other things Réaumur's insistence on a one-point calibration and the fact that the quality of French wine varied from year to year, led to all sorts of complications. So although the French gave this thermometer a good try for over a century, they eventually gave it up and drank their wine instead.

While these developments were taking place in warmer climes, Swedish astronomer Anders Celsius tramped his snowy land with ten cold fingers and ten cold toes advocating a 100 division scale (centigrade) by continually proclaiming "water boils at 0°, water freezes at 100°, boils at 0°, freezes at 100." By happenstance however, Anders' intimate friend, Linnaeus, the great botanist, was left-handed. Because of this he kept snatching the wrong end of the instrument and using it upside down, and kept reading 0° for the freezing point and 100 for the boiling point, and thus recommended this scale.

To confuse the issue further, there were other strong claimants for the honor of inventing this scale; nevertheless, in 1948 the 9th General Conference of Weights and Measures decided that it knew enough, it dismissed all the others, and ruled that what had been known as degrees Centigrade would henceforth and forever more be known as degrees Celsius. And so Celsius' name will be with us forever, and all because he chose his parents wisely. Had Linnaeus' name been Clinnaeus we might today be talking of degrees Clinnaeus instead of degrees Celsius.

A.   CALIBRATION CONFUSION

Thermometers evolved into three broad types, as shown in Figure 25–1, and as users demanded more precision all sorts of problems cropped up. For example, should melting ice or freezing water be one of the calibration points? In practice they differ! Should boiling water or condensing steam be the other? Also, the zero point of a given thermometer slowly and continually changed with time. In mercury thermometers it crept upward; in alcohol thermometers it slid downward. Was this due to the aging of glass, the slow decomposition of the fluids, or what? And these changes continued for ten, twenty, thirty years!! Also, how did these scientists explain the small periodic variations according to season?

Probably the most serious problem was that when one type of thermometer read a temperature halfway between calibration points, the others did not because of changing coefficients of expansion of fluids. This is illustrated in Figure 25–1. In this case, which thermometer read the true midpoint temperature—which to trust? Was the selection of equal intervals of temperature an arbitrary matter, or was there a rational way for doing this? These difficulties puzzled scientists for quite a while.

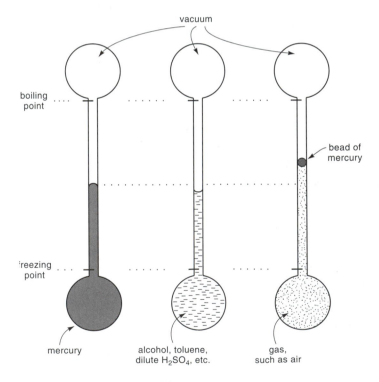

Figure 25–1.

In the 1800s there was much concern about developing a rational temperature scale. In 1847 Regnault pinpointed the problem by stating:

> We give the name *thermometer* to instruments intended to measure the varia-
> tion of the quantity of heat in a body . . . A perfect thermometer would be one
> for which the addition of equal quantities of heat always produces equal ex-
> pansions . . . Unfortunately this is not so for real substances.

Just a year later William Thomson, later Lord Kelvin, magically devised
just such a temperature scale based upon the concept of the ideal reversible heat
engine of Sadi Carnot. We discussed and developed this temperature scale in
Chapter 18. On this basis we now commonly use two scales; the Kelvin scale corresponding to degrees Celsius and the Rankine scale corresponding to degrees
Fahrenheit. These are sketched in Figure 25–2.

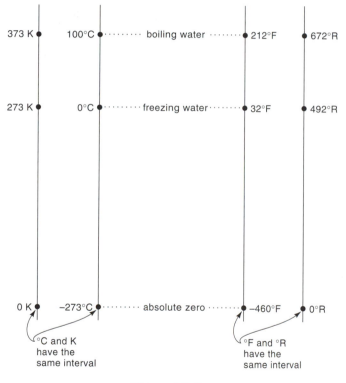

Figure 25–2.

B. TEMPERATURE TODAY

So here is how things stand today. First, the lowest possible temperature imagin-
able, the absolute zero, has been invented. This is where our temperature scales
should start from, and this is where our absolute scales (Kelvin and °Rankine) in
fact do. Secondly, we have a rational way for choosing equal intervals of temperature.

We just have one question left. Starting from this cold, cold zero point, why pick our unit of temperature the way we do? Why base it on 100 intervals or 180 intervals between the freezing and boiling point of water? Why pick water, why not googliox? Isn't there a more reasonable way of selecting our unit of temperature?

As Regnault long ago pointed out, the temperature measures the "quantity of heat" in a body, or in today's language, its "thermal energy." So why not measure temperature directly as *energy per unit quantity of material*? J. C. Georgian (*Nature, 201,* 695 (1964); *203,* 1158 (1964); *207,* 1285 (1965)) strongly urges that we adopt such a scale choosing the ideal gas as measuring instrument. The reason for this is that the energy of any ideal gas is proportional to its absolute temperature. So measure its energy and you've got its temperature.

With this ideal-gas-energy temperature scale, or IGE scale, water freezes at

$$T = pv = (101\ 325\ \text{Pa})(0.0224\ \frac{\text{m}^3}{\text{mol}}) = 2270\ \frac{\text{J}}{\text{mol}}$$

and boils at

$$T = 2270 \left(\frac{373.15}{273.15} \right) = 3100\ \frac{\text{J}}{\text{mol}}$$

Figure 25–3 compares the various absolute temperature scales following this proposal.

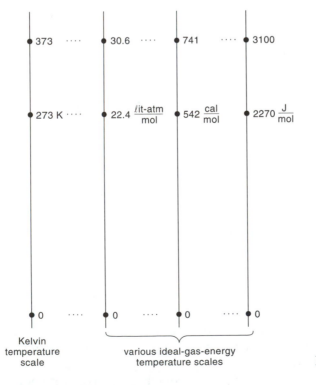

Kelvin temperature scale

various ideal-gas-energy temperature scales

Figure 25–3.

It may seem awkward and foreign to talk of water freezing at 2270°X, boiling at 3100°X, and of having a mild fever of 2600°X (where X would then honor some famous scientist whose name does not start with the letters C, F, K or R).[1] However, with such a scale a number of simplifications occur naturally. In particular, it would forever banish one conversion factor from our lives (all thermo students would cheer this), that miserable gas constant. Thus, for a mole of ideal gas

and

$$pv = RT \text{ would become } \quad pv = T \left[\frac{J}{mol} \right]$$

$$c_p - c_v = R \text{ would become } \quad c_p - c_v = 1 \, [-]$$

Also, for a mole of any substance

$$c_p \text{ and s would be dimensionless}$$

This temperature scale should help clarify the meaning of the concept of entropy because it shows that the entropy measures the fraction of energy of a substance which is in random motion. It would also simplify and make more meaningful the study of thermo.

C. THE MEANING OF ENTROPY, THE THIRD LAW

When energy is removed from a material, its temperature drops and the movement of its molecules slows (reduced kinetic energy). Think of water. As steam the molecules buzz about and collide with each other at about 400 m/s. As liquid the molecules slide about, always next to each other, moving much, much slower than the gas. When more energy is removed the liquid turns into ice, the molecules do not slide about but stay in fixed positions, vibrating about these positions.

Now the third law of thermo says that at absolute zero in temperature, or 0 K, even the vibration of crystalline solids ceases, and everything, even on the molecular or atomic level, is frozen. This means that everything is known about every molecule and every atom—where each atom is located, and where it is going (nowhere). We have complete information about the system; there is no un-

---

[1]Why not choose "G" to recognize John Gregorian who first proposed this rational temperature scale?

certainty in our knowledge, no randomness. Referring to the previous chapter on information theory we can say that at 0 K

Regarding information: I is maximum

Regarding the uncertainty: S is minimum, in fact zero

This, then, is the *third law of thermo*. It tells that S = 0 at 0 K.

No one has been able to get down to absolute zero in temperature; however, researchers (see *Science News, 146,* 175 (1994)) have recently gotten to as close as one millionth of a degree from 0 K. At that temperature atoms do not fly about at hundreds of meters per second, but lazily float about at an average 1 cm/s. Whatever the temperature may be (above 0 K) this leads to uncertainty in our knowledge about the molecules.

This behavior is reflected in the IGE temperature scale, which tells directly, in dimensionless measure,

- at 0 K   S = (the energy in random motion) = 0
- above 0 K S = (the energy in the form of random motion of molecules)>0.
- at very high temperature the molecules buzz about frantically and hysterically so the entropy is very large.

The ideal-gas-energy is a rational temperature scale based on the principles of thermo. I wonder whether a change to such a temperature scale could ever receive consideration, or is science too big with too much inertia? Time will tell.

DIMENSIONS, UNITS AND CONVERSIONS AND THERMO PROPERTIES OF H_2O AND HFC-134A

A. NEWTON'S LAW

at Earth's surface $a = g = 9.806$ m/s^2

$$F = \frac{ma}{g_c}$$

conversion factor $g_c = 1\ \dfrac{\text{kg·m}}{\text{s}^2\text{·N}}$

B. LENGTH

10^{10}	10^6	39.370	3.280 84	1	0.000 6214
angstrom	micron	inch	foot	meter	mile

C. VOLUME

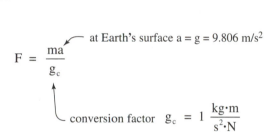

61 024	35 195	33 814	1000	264.2	219.0
in^3	UK fl oz	US fl oz	lit	US gal	Imp. gal

35.315	6.290	4.80	1
ft^3	bbl (oil)	drum	m^3

42 US gal   55 US gal

Also

768	256	133.23	128	8	4	1
US tsp	US tbsp	UK fl oz	US fl oz	US pint	US qt	US gal

Also

160	10	1
UK fl oz	UK pint	UK gal

D. MASS

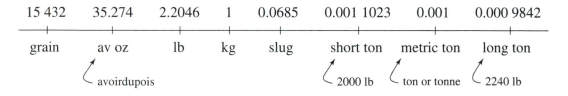

15 432	35.274	2.2046	1	0.0685	0.001 1023	0.001	0.000 9842
grain	av oz	lb	kg	slug	short ton	metric ton	long ton
	avoirdupois				2000 lb	ton or tonne	2240 lb

E. PRESSURE

The Pascal : $1 \text{ Pa} = 1\dfrac{N}{m^2} = 1\dfrac{kg}{m\cdot s^2} = 10\dfrac{dyne}{cm^2}$

$1 \text{ atm} = 760 \text{ mm Hg} = 760 \text{ torr} = 14.696 \dfrac{lb_f}{in^2}$

$= 29.92 \text{ in Hg} = 33.93 \text{ ft } H_2O = 407.189 \text{ in } H_2O$

$= 101 \; 325 \text{ Pa}$

$1 \text{ bar} = 10^5 \text{ Pa}$, close to 1 atm, sometimes called a technical atmosphere

$1 \text{ in } H_2O = 248.84 \text{ Pa} \cong 250 \text{ Pa}$

F. WORK, HEAT, AND ENERGY

The joule : $1 \text{ Joule} = 1 \text{ N}\cdot\text{m} = 1 \dfrac{kg\cdot m}{s^2}$

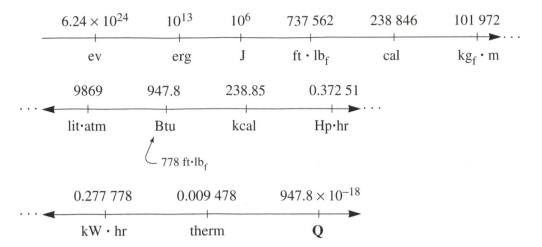

G. POWER

The watt: $1 \text{ W} = 1 \text{ J/s}$

1000	1	1.341
W	kW	Hp

H. MOLECULAR WEIGHT OR MOLECULAR MASS

$$\text{In SI units : } (\overline{mw})_{0_2} = 0.032 \frac{\text{kg}}{\text{mol}}$$

$$(\overline{mw})_{air} = 0.0289 \frac{\text{kg}}{\text{mol}}$$

I. IDEAL GAS

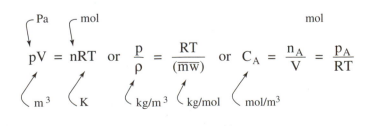

$$\text{the gas constant} \quad R = 8.314 \ \frac{J}{mol \cdot K} = 1.987 \ \frac{cal}{mol \cdot K} = 0.7302 \ \frac{ft^3 \cdot atm}{lb \ mol \cdot K}$$

$$= 0.082 \ 06 \ \frac{lit \cdot atm}{mol \cdot K} = 1.987 \ \frac{Btu}{lb \ mol \cdot K} = 8.314 \ \frac{Pa \cdot m^3}{mol \cdot K}$$

J. WATER AND STEAM—SATURATION PROPERTIES: CONSTANT PRESSURE TABLE

Pressure (kPa)	Temp. (°C)	Specific volume (m³/kg)		Internal energy (kJ/kg)		Enthalpy (kJ/kg)		Entropy (kJ/kg.K)	
		sat. liq.	sat. vap.	sat. liq.	sat. vap.	sat. liq.	sat. vap.	sat. liq.	sat. vap.
0.6113	0.01	0.001 000	206.14	.00	2375.3	.01	2501.4	.0000	9.1562
1.0	6.98	0.001 000	129.21	29.30	2385.0	29.30	2514.2	.1059	8.9756
2.0	17.50	0.001 001	67.00	73.48	2399.5	73.48	2533.5	.2607	8.7237
3.0	24.08	0.001 003	45.67	101.04	2408.5	101.05	2545.5	.3545	8.5776
5.0	32.88	0.001 005	28.19	137.81	2420.5	137.82	2561.5	.4764	8.3951
10	45.81	0.001 010	14.67	191.82	2437.9	191.83	2584.7	.6493	8.1502
20	60.06	0.001 017	7.649	251.38	2456.7	251.40	2609.7	.8320	7.9085
30	69.10	0.001 022	5.229	289.20	2468.4	289.23	2625.3	.9439	7.7686
40	75.87	0.001 027	3.993	317.53	2477.0	317.58	2636.8	1.0259	7.6700
50	81.33	0.001 030	3.240	340.44	2483.9	340.49	2645.9	1.0910	7.5939
75	91.78	0.001 037	2.217	384.31	2496.7	384.39	2663.0	1.2130	7.4564
MPa									
0.10	99.63	0.001 043	1.6940	417.36	2506.1	417.46	2675.5	1.3026	7.3594
0.15	111.37	0.001 053	1.1593	466.94	2519.7	467.11	2693.6	1.4336	7.2233
0.20	120.23	0.001 061	0.8857	504.49	2529.5	504.70	2706.7	1.5301	7.1271
0.25	127.44	0.001 067	0.7187	535.10	2537.2	535.37	2716.9	1.6072	7.0527
0.30	133.55	0.001 073	0.6058	561.15	2543.6	561.47	2725.3	1.6718	6.9919
0.40	143.63	0.001 084	0.4625	604.31	2553.6	604.74	2738.6	1.7766	6.8959
0.50	151.86	0.001 093	0.3749	639.68	2561.2	640.23	2748.7	1.8607	6.8213
0.70	164.97	0.001 108	0.2729	696.44	2572.5	697.22	2763.5	1.9922	6.7080
0.75	167.78	0.001 112	0.2556	708.64	2574.7	709.47	2766.4	2.0200	6.6847
1.00	179.91	0.001 127	0.194 44	761.68	2583.6	762.81	2778.1	2.1387	6.5865
1.50	198.32	0.001 154	0.131 77	843.16	2594.5	844.89	2792.2	2.3150	6.4448
2.00	212.42	0.001 177	0.099 63	906.44	2600.3	908.79	2799.5	2.4474	6.3409
3.0	233.90	0.001 217	0.066 68	1004.78	2604.1	1008.42	2804.2	2.6457	6.1869
5	263.99	0.001 286	0.039 44	1147.81	2597.1	1154.23	2794.3	2.9202	5.9734
7	285.88	0.001 351	0.027 37	1257.55	2580.5	1267.00	2772.1	3.1211	5.8133
10	311.06	0.001 452	0.018 026	1393.04	2544.4	1407.56	2724.7	3.3596	5.6141
15	342.24	0.001 658	0.010 337	1585.6	2455.5	1610.5	2610.5	3.6848	5.3098
20	365.81	0.002 036	0.005 834	1785.6	2293.0	1826.3	2409.7	4.0139	4.9269
22.09	374.14	0.003 155	0.003 155	2029.6	2029.6	2099.3	2099.3	4.4298	4.4298

Data from J. H. Keenan, F. G. Keys, P. G. Hill and J. G. Moore, *Steam Tables*, © 1969, John Wiley and Sons; abridged from B. G. Kyle, *Chemical and Process Thermodynamics*, 2nd ed., Prentice Hall, 1992, with permission.

K. WATER AND STEAM—SATURATION PROPERTIES: CONSTANT TEMPERATURE TABLE

Temp. (°C)	Pressure (kPa)	Specific volume (m³/kg)		Internal energy (kJ/kg)		Enthalpy (kJ/kg)		Entropy (kJ/kg.K)	
		sat. liq.	sat. vap.	sat. liq.	sat. vap.	sat. liq.	sat. vap.	sat. liq.	sat. vap.
0.01	0.6113	0.001 000	206.14	.00	2375.3	.01	2501.4	.0000	9.1562
5	0.8721	0.001 000	147.12	20.97	2382.3	20.98	2510.6	.0761	9.0257
10	1.2276	0.001 000	106.38	42.00	2389.2	42.01	2519.8	.1510	8.9008
15	1.7051	0.001 001	77.93	62.99	2396.1	62.99	2528.9	.2245	8.7814
20	2.339	0.001 002	57.79	83.95	2402.9	83.96	2538.1	.2966	8.6672
25	3.169	0.001 003	43.36	104.88	2409.8	104.89	2547.2	.3764	8.5580
30	4.246	0.001 004	32.89	125.78	2416.6	125.79	2556.3	.4369	8.4533
35	5.628	0.001 006	25.22	146.67	2423.4	146.68	2565.3	.5053	8.3531
40	7.384	0.001 008	19.52	167.56	2430.1	167.57	2574.3	.5725	8.2570
45	9.593	0.001 010	15.26	188.44	2436.8	188.45	2583.2	.6387	8.1648
50	12.349	0.001 012	12.03	209.32	2443.5	209.33	2592.1	.7038	8.0763
55	15.758	0.001 015	9.568	230.21	2450.1	230.23	2600.9	.7679	7.9913
60	19.940	0.001 017	7.671	251.11	2456.6	251.13	2609.6	.8312	7.9096
65	25.03	0.001 020	6.197	272.02	2463.1	272.06	2618.3	.8935	7.8310
70	31.19	0.001 023	5.042	292.95	2469.6	292.98	2626.8	.9549	7.7553
75	38.58	0.001 026	4.131	313.90	2475.9	313.93	2635.3	1.0155	7.6824
80	47.39	0.001 029	3.407	334.86	2482.2	334.91	2643.7	1.0753	7.6122
85	57.83	0.001 033	2.828	355.84	2488.4	355.90	2651.9	1.1343	7.5445
90	70.14	0.001 036	2.361	376.85	2494.5	376.92	2660.1	1.1925	7.4791
95	84.55	0.001 040	1.982	397.88	2500.6	397.96	2668.1	1.2500	7.4159
100	0.101 35	0.001 044	1.6729	418.94	2506.5	419.04	2676.1	1.3069	7.3549
110	0.143 27	0.001 052	1.2102	461.14	2518.1	461.30	2691.5	1.4185	7.2387
120	0.198 53	0.001 060	0.8919	503.50	2529.3	503.71	2706.3	1.5276	7.1296
130	0.2701	0.001 070	0.6685	546.02	2539.9	546.31	2720.5	1.6344	7.0269
140	0.3613	0.001 080	0.5089	588.74	2550.0	589.13	2733.9	1.7391	6.9299
150	0.4758	0.001 091	0.3928	631.68	2559.5	632.20	2746.5	1.8418	6.8379
160	0.6178	0.001 102	0.3071	674.87	2568.4	675.55	2758.1	1.9427	6.7502
170	0.7917	0.001 114	0.2428	718.33	2576.5	719.21	2768.7	2.0419	6.6663
180	1.0021	0.001 127	0.194 05	762.09	2583.7	763.22	2778.2	2.1396	6.5857
190	1.2544	0.001 141	0.156 54	806.19	2590.0	807.62	2786.4	2.2359	6.5079
200	1.5538	0.001 157	0.127 36	850.65	2595.3	852.45	2793.2	2.3309	6.4323
210	1.9062	0.001 173	0.104 41	895.53	2599.5	897.76	2798.5	2.4248	6.3585
220	2.318	0.001 190	0.086 19	940.87	2602.4	943.62	2802.1	2.5178	6.2861
230	2.795	0.001 209	0.071 58	986.74	2603.9	990.12	2804.0	2.6099	6.2146
240	3.344	0.001 229	0.059 76	1033.21	2604.0	1037.32	2803.8	2.7015	6.1437
250	3.973	0.001 251	0.050 13	1080.39	2602.4	1085.36	2801.5	2.7927	6.0730
260	4.688	0.001 276	0.042 21	1128.39	2599.0	1134.37	2796.9	2.8838	6.0019
270	5.499	0.001 302	0.035 64	1177.36	2593.7	1184.51	2789.7	2.9751	5.9301
280	6.412	0.001 332	0.030 17	1227.46	2586.1	1235.99	2779.6	3.0668	5.8571
290	7.436	0.001 366	0.025 57	1278.92	2576.0	1289.07	2766.2	3.1594	5.7821

K.   WATER AND STEAM—SATURATION PROPERTIES: CONSTANT TEMPERATURE TABLE (CONTINUED)

Temp. (°C)	Pressure (kPa)	Specific volume (m³/kg)		Internal energy (kJ/kg)		Enthalpy (kJ/kg)		Entropy (kJ/kg.K)	
		sat. liq.	sat. vap.	sat. liq.	sat. vap.	sat. liq.	sat. vap.	sat. liq.	sat. vap.
300	8.581	0.001 404	0.021 67	1332.0	2563.0	1344.0	2749.0	3.2534	5.7045
310	9.856	0.001 447	0.018 350	1387.1	2546.4	1401.3	2727.3	3.3493	5.6230
320	11.274	0.001 499	0.015 488	1444.6	2525.5	1461.5	2700.1	3.4480	5.5362
330	12.845	0.001 561	0.012 996	1505.3	2498.9	1525.3	2665.9	3.5507	5.4417
340	14.586	0.001 638	0.010 797	1570.3	2464.6	1594.2	2622.0	3.6594	5.3357
350	16.513	0.001 740	0.008 813	1641.9	2418.4	1670.6	2563.9	3.7777	5.2112
360	18.651	0.001 893	0.006 945	1725.2	2351.5	1760.5	2481.0	3.9147	5.0526
370	21.03	0.002 213	0.004 925	1844.0	2228.5	1890.5	2332.1	4.1106	4.7971
374.14	22.09	0.003 155	0.003 155	2029.6	2029.6	2099.3	2099.3	4.4298	4.4298

Data from J. H. Keenan, F. G. Keys, P. G. Hill, and J. G. Moore, *Steam Tables,* © 1969, John Wiley and Sons; abridged from G. J. VanWylen and R. E. Sonntag, *Fundamentals of Classical Thermodynamics,* 2nd ed., S.I. Version, John Wiley and Sons, New York, 1976, by permission from John Wiley and Sons, Inc.

L.   STEAM—SUPERHEATED PROPERTIES: CONSTANT PRESSURE TABLE

T	p = .010 MPa (45.81)				p = .050 MPa (81.33)				p = .10 MPa (99.63)			
	v	u	h	s	v	u	h	s	v	u	h	s
Sat.	14.674	2437.9	2584.7	8.1502	3.240	2483.9	2645.9	7.5939	1.6940	2506.1	2675.5	7.3594
50	14.869	2443.9	2592.6	8.1749								
100	17.196	2515.5	2687.5	8.4479	3.418	2511.6	2682.5	7.6947	1.6958	2506.7	2676.2	7.3614
150	19.512	2587.9	2783.0	8.6882	3.889	2585.6	2780.1	7.9401	1.9364	2582.8	2776.4	7.6134
200	21.825	2661.3	2879.5	8.9038	4.356	2659.9	2877.7	8.1580	2.172	2658.1	2875.3	7.8343
250	24.136	2736.0	2977.3	9.1002	4.820	2735.0	2976.0	8.3556	2.406	2733.7	2974.3	8.0333
300	26.445	2812.1	3076.5	9.2813	5.284	2811.3	3075.5	8.5373	2.639	2810.4	3074.3	8.2158
400	31.063	2968.9	3279.6	9.6077	6.209	2968.5	3278.9	8.8642	3.103	2967.9	3278.2	8.5435
500	35.679	3132.3	3489.1	9.8978	7.134	3132.0	3488.7	9.1546	3.565	3131.6	3488.1	8.8342
600	40.295	3302.5	3705.4	10.1608	8.057	3302.2	3705.1	9.4178	4.028	3301.9	3704.7	9.0976
	p = .20 MPa (120.23)				p = .30 MPa (133.55)				p = .40 MPa (143.63)			
Sat.	.8857	2529.5	2706.7	7.1272	.6058	2543.6	2725.3	6.9919	.4625	2553.6	2738.6	6.8959
150	.9596	2576.9	2768.8	7.2795	.6339	2570.8	2761.0	7.0778	.4708	2564.5	2752.8	6.9299
200	1.0803	2654.4	2870.5	7.5066	.7163	2650.7	2865.6	7.3115	.5342	2646.8	2860.5	7.1706
250	1.1988	2731.2	2971.0	7.7086	.7964	2728.7	2967.6	7.5166	.5951	2726.1	2964.2	7.3789
300	1.3162	2808.6	3071.8	7.8926	.8753	2806.7	3069.3	7.7022	.6548	2804.8	3066.8	7.5662
400	1.5493	2966.7	3276.6	8.2218	1.0315	2965.6	3275.0	8.0330	.7726	2964.4	3273.4	7.8985
500	1.7814	3130.8	3487.1	8.5133	1.1867	3130.0	3486.0	8.3251	.8893	3129.2	3484.9	8.1913
600	2.013	3301.4	3704.0	8.7770	1.3414	3300.8	3703.2	8.5892	1.0055	3300.2	3702.4	8.4558

	p = .50 MPa (151.86)				p = .60 MPa (158.85)				p = 1.00 MPa (179.91)			
T	v	u	h	s	v	u	h	s	v	u	h	s
Sat.	.3749	2561.2	2748.7	6.8213	.3157	2567.4	2756.8	6.7600	.194 44	2583.6	2778.1	6.5865
200	.4249	2642.9	2855.4	7.0592	.3520	2638.9	2850.1	6.9665	.2060	2621.9	2827.9	6.6940
250	.4744	2723.5	2960.7	7.2709	.3938	2720.9	2957.2	7.1816	.2327	2709.9	2942.6	6.9247
300	.5226	2802.9	3064.2	7.4599	.4344	2801.0	3061.6	7.3724	.2579	2793.2	3051.2	7.1229
350	.5701	2882.6	3167.7	7.6329	.4742	2881.2	3165.7	7.5464	.2825	2875.2	3157.7	7.3011
400	.6173	2963.2	3271.9	7.7938	.5137	2962.1	3270.3	7.7079	.3066	2957.3	3263.9	7.4651
500	.7109	3128.4	3483.9	8.0873	.5920	3127.6	3482.8	8.0021	.3541	3124.4	3478.5	7.7622
600	.8041	3299.6	3701.1	7.3522	.6697	3299.1	3700.9	8.2674	.4011	3296.8	3697.9	8.0290
700	.8969	3477.5	3925.9	8.5952	.7472	3477.0	3925.3	8.5107	.4478	3475.3	3923.1	8.2731
800	.9896	3662.1	4156.9	8.8211	.8245	3661.8	4156.5	8.7367	.4943	3660.4	4154.7	8.4996
900	1.0822	3835.6	4394.7	9.0329	.9017	3853.4	4394.4	8.9486	.5407	3852.2	4392.9	8.7118
1000	1.1747	4051.8	4639.1	9.2328	.9788	4051.5	4638.8	9.1485	.5871	4050.5	4637.6	8.9119

	p = 1.40 MPa (195.07)				p = 2.00 MPa (212.42)				p = 3.00 MPa (233.90)			
Sat.	.140 84	2592.8	2790.0	6.4693	.099 63	2600.3	2799.5	6.3409	.066 68	2604.1	2804.2	6.1869
200	.143 02	2603.1	2803.3	6.4975								
250	.163 50	2698.3	2927.2	6.7467	.111 44	2679.6	2902.5	6.5453	.070 58	2644.0	2855.8	6.2872
300	.182 28	2785.2	3040.4	6.9534	.125 47	2772.6	3023.5	6.7664	.081 14	2750.1	2993.5	6.5390
350	.2003	2869.2	3149.5	7.1360	.138 57	2859.8	3137.0	6.9563	.090 53	2843.7	3115.3	6.7428
400	.2178	2952.5	3257.5	7.3026	.151 20	2945.2	3247.6	7.1271	.099 36	2932.8	3230.9	6.9212
									.107 87	3020.4	3344.0	7.0834
500	.2521	3121.1	3474.1	7.6027	.175 68	3116.2	3467.6	7.4317	.116 19	3108.0	3456.5	7.2338
600	.2860	3294.4	3694.8	7.8710	.199 60	3290.9	3690.1	7.7024	.132 43	3285.0	3682.3	7.5085
700	.3195	3473.6	3920.8	8.1160	.2232	3470.9	3917.4	7.9487	.148 38	3466.5	3911.7	7.7571
800	.3528	3659.0	4153.0	8.3431	.2467	3657.0	4150.3	8.1765	.164 14	3653.5	4145.9	7.9862
900	.3861	3851.1	4391.5	8.5556	.2700	3849.3	4389.4	8.3895	.179 80	3846.5	4385.9	8.1999
1000	.4192	4049.5	4636.4	8.7559	.2933	4048.0	4634.6	8.5901	.195 41	4045.4	4631.6	8.4009

	p = 4.0 MPa (250.40)				p = 6.0 MPa (275.64)				p = 8.0 MPa (295.06)			
Sat.	.049 78	2602.3	2801.4	6.0701	.032 44	2589.7	2784.3	5.8892	.023 52	2569.8	2758.0	5.7432
300	.058 84	2725.3	2960.7	6.3615	.036 16	2667.2	2884.2	6.0674	.024 26	2590.9	2785.0	5.7906
350	.066 45	2826.7	3092.5	6.5821	.042 23	2789.6	3043.0	6.3335	.029 95	2747.7	2987.3	6.1301
400	.073 41	2919.9	3213.6	6.7690	.047 39	2892.9	3177.2	6.5408	.034 32	2863.8	3138.3	6.3634
450	.080 02	3010.2	3330.3	6.9363	.052 14	2988.9	3301.8	6.7193	.038 17	2966.7	3272.0	6.5551
500	.086 43	3099.5	3445.3	7.0901	.056 65	3082.2	3422.2	6.8803	.041 75	3064.3	3398.3	6.7240
550					.061 01	3174.6	3540.6	7.0288	.045 16	3159.8	3521.0	6.8778
600	.098 85	3279.1	3674.4	7.3688	.065 25	3266.9	3658.4	7.1677	.048 45	3254.4	3642.0	7.0206
700	.110 95	3462.1	3905.9	7.6198	.073 52	3453.1	3894.2	7.4234	.054 81	3443.9	3882.4	7.2812
800	.122 87	3650.0	4141.5	7.8502	.081 60	3643.1	4132.7	7.6566	.060 97	3636.0	4123.8	7.5173
900	.134 69	3843.6	4382.3	8.0647	.089 58	3837.8	4375.3	7.8727	.067 02	3832.1	4368.3	7.7351
1000	.146 45	4042.9	4628.7	8.2662	.097 49	4037.8	4622.7	8.0751	.073 01	4032.8	4616.9	7.9384

Data from J. H. Keenan, F. G. Keys, P. G. Hill, and J. G. Moore, *Steam Tables,* © 1969, John Wiley and Sons; abridged from B. G. Kyle, *Chemical and Process Thermodynamics,* 2nd ed., Prentice Hall, 1992, with permission.

M. HFC-134A SATURATION PROPERTIES

Temp. (°C)	Pressure (kPa)		Density (kg/m³)	Enthalpy (kJ/kg)	Entropy (kJ/kg·K)	Temp. (°C)	Pressure (kPa)		Density (kg/m³)	Enthalpy (kJ/kg)	Entropy (kJ/kg·K)
−100	0.566	liq.	1580.49	77.268	0.4448	5	349.868	liq.	1276.74	206.756	1.0244
		vap.	0.040	337.199	1.9460			vap.	17.140	401.697	1.7252
−95	0.947	liq.	1566.80	83.033	0.4776	10	414.919	liq.	1259.77	213.593	1.0485
		vap.	0.065	340.146	1.9209			vap.	20.236	404.532	1.7229
−90	1.532	liq.	1553.10	88.792	0.5095	15	488.801	liq.	1242.33	220.514	1.0726
		vap.	0.103	343.132	1.8982			vap.	23.770	407.300	1.7208
−85	2.407	liq.	1539.41	94.565	0.5406	20	572.259	liq.	1224.38	227.526	1.0964
		vap.	0.158	346.155	1.8778			vap.	27.791	409.993	1.7189
−80	3.679	liq.	1525.71	100.367	0.5710	25	666.063	liq.	1205.86	234.634	1.1202
		vap.	0.235	349.209	1.8594			vap.	32.359	412.600	1.7171
−75	5.482	liq.	1512.01	106.206	0.6009	30	771.005	liq.	1186.69	241.846	1.1439
		vap.	0.341	352.292	1.8428			vap.	37.540	415.109	1.7155
−70	7.980	liq.	1498.29	112.087	0.6302	35	887.907	liq.	1166.81	249.170	1.1676
		vap.	0.486	355.399	1.8279			vap.	43.413	417.506	1.7138
−65	11.371	liq.	1484.54	118.014	0.6590	40	1017.616	liq.	1146.11	256.617	1.1912
		vap.	0.677	358.525	1.8144			vap.	50.072	419.775	1.7122
−60	15.887	liq.	1470.73	123.989	0.6873	45	1161.013	liq.	1124.49	264.198	1.2148
		vap.	0.926	361.667	1.8024			vap.	57.630	421.897	1.7105
−55	21.797	liq.	1456.86	130.014	0.7152	50	1319.017	liq.	1101.80	271.926	1.2384
		vap.	1.245	364.818	1.7916			vap.	66.225	423.845	1.7086
−50	29.406	liq.	1442.89	136.091	0.7428	55	1492.592	liq.	1077.87	279.821	1.2622
		vap.	1.648	367.975	1.7819			vap.	76.035	425.590	1.7064
−45	39.059	liq.	1428.82	142.219	0.7699	60	1682.762	liq.	1052.47	287.905	1.2861
		vap.	2.149	371.134	1.7732			vap.	87.287	427.091	1.7039
−40	51.139	liq.	1414.61	148.401	0.7967	65	1890.623	liq.	1025.28	296.209	1.3102
		vap.	2.767	374.288	1.7655			vap.	100.283	428.296	1.7009
−35	66.065	liq.	1400.25	154.638	0.8231	70	2117.366	liq.	995.91	304.772	1.3347
		vap.	3.518	377.434	1.7586			vap.	115.442	429.132	1.6971
−30	84.295	liq.	1385.71	160.932	0.8492	75	2364.313	liq.	963.73	313.652	1.3597
		vap.	4.424	380.566	1.7525			vap.	133.373	429.496	1.6924
−25	106.320	liq.	1370.97	167.283	0.8750	80	2632.970	liq.	927.84	322.936	1.3854
		vap.	5.504	383.681	1.7470			vap.	155.010	429.230	1.6863
−20	132.668	liq.	1356.00	173.695	0.9005	85	2925.109	liq.	886.70	332.764	1.4121
		vap.	6.784	386.773	1.7422			vap.	181.929	428.076	1.6782
−15	163.899	liq.	1340.77	180.170	0.9257	90	3242.934	liq.	837.34	343.414	1.4406
		vap.	8.288	389.836	1.7379			vap.	217.162	425.547	1.6668
−10	200.601	liq.	1325.27	186.711	0.9507	95	3589.451	liq.	772.35	355.577	1.4727
		vap.	10.044	392.866	1.7341			vap.	268.255	420.471	1.6489
−5	243.394	liq.	1309.45	193.319	0.9755	100	3969.943	liq.	651.44	373.174	1.5187
		vap.	12.082	395.857	1.7308			vap.	375.503	406.956	1.6092
0	**292.925**	**liq.**	**1293.28**	**200.000**	**1.0000**						
		vap.	**14.435**	**398.803**	**1.7278**						

N. HFC-134A SUPERHEATED VAPOR: CONSTANT PRESSURE TABLES

Temp. (°C)	Pressure = 50kPa p, (kg/m³)	h, (kJ/kg)	s, (kJ/kg·K)		Temp. (°C)	Pressure = 100kPa p, (kg/m³)	h, (kJ/kg)	s, (kJ/kg·K)
−40.43	1408.451	147.9	0.7944	Sat. Liquid	−26.34	1369.863	165.6	0.8681
−40.43	2.709	374.0	1.7661	Sat. Vapor	−26.34	5.195	382.8	1.7484
−40	2.704	374.3	1.7675		−20	5.043	387.9	1.7685
−30	2.581	381.8	1.7990		−10	4.823	395.8	1.7994
−20	2.471	389.5	1.8297		0	4.624	404.0	1.8297
−10	2.370	397.3	1.8599		10	4.444	412.2	1.8593
0	2.278	405.2	1.8895		20	4.279	420.6	1.8883
10	2.194	413.3	1.9186		30	4.127	429.1	1.9169
20	2.116	421.5	1.9473		40	3.986	437.7	1.9450
30	2.043	430.0	1.9755		50	3.855	446.6	1.9727
40	1.976	438.5	2.0034		60	3.733	455.5	2.0001
50	1.913	447.3	2.0309		70	3.620	464.7	2.0270
60	1.854	456.2	2.0580		80	3.513	473.9	2.0537

Temp. (°C)	Pressure = 200kPa				Temp.	Pressure = 300kPa		
−10.08	1333.333	186.8	0.9503	Sat. Liquid	0.66	1298.701	200.9	1.0032
−10.08	10.015	392.8	1.7342	Sat. Vapor	0.66	14.771	399.2	1.7275
−20	1356.195	173.7	0.9004		10	14.101	407.5	1.7573
−10	10.011	392.9	1.7344		20	13.469	416.4	1.7883
0	9.544	401.4	1.7661		30	12.907	425.4	1.8183
10	9.130	409.9	1.7968		40	12.402	434.4	1.8477
20	8.759	418.5	1.8267		50	11.944	443.6	1.8764
30	8.422	427.3	1.8560		60	11.523	452.8	1.9045
40	8.114	436.1	1.8847		70	11.138	462.1	1.9322
50	7.833	445.1	1.9129		80	10.781	471.6	1.9594
60	7.572	454.2	1.9406		90	10.448	481.2	1.9862
70	7.330	463.4	1.9679		100	10.139	490.9	2.0126
80	7.105	472.8	1.9948		110	9.849	500.8	2.0387

Temp. (°C)	Pressure = 500kPa				Temp.	Pressure = 600kPa		
15.71	1234.568	221.5	1.0759	Sat. Liquid	21.54	1219.512	229.7	1.1038
15.71	24.307	407.7	1.7205	Sat. Vapor	21.54	29.137	410.8	1.7183
20	23.737	411.8	1.7347		30	27.781	419.2	1.7463
30	22.551	421.3	1.7667		40	26.404	428.9	1.7780
40	21.523	430.8	1.7975		50	25.209	438.7	1.8086
50	20.615	440.4	1.8274		60	24.155	448.4	1.8382
60	19.803	449.9	1.8565		70	23.213	458.1	1.8671
70	19.070	459.5	1.8849		80	22.361	468.0	1.8953
80	18.402	469.2	1.9128		90	21.585	477.9	1.9229
90	17.791	479.0	1.9401		100	20.873	487.8	1.9501
100	17.224	489.0	1.9670		110	20.221	497.9	1.9767
110	16.698	498.9	1.9934		120	19.608	508.1	2.0030
120	16.210	509.0	2.0195		130	19.041	518.4	2.0288

Temp. (°C)	p, (kg/m³)	Pressure = 800kPa h, (kJ/kg)	s, (kJ/kg·K)		Temp. (°C)	p, (kg/m³)	Pressure = 1000kPa h, (kJ/kg)	s, (kJ/kg·K)
31.29	1176.471	243.7	1.1500	Sat. Liquid	39.35	1149.425	255.6	1.1881
31.29	38.986	415.7	1.7150	Sat. Vapor	39.35	49.158	419.5	1.7124
40	36.968	424.8	1.7445		40	48.932	420.2	1.7147
50	35.017	435.1	1.7767		50	45.849	431.2	1.7491
60	33.352	445.2	1.8076		60	43.330	441.8	1.7816
70	31.897	455.3	1.8374		70	41.201	452.3	1.8126
80	30.611	465.4	1.8664		80	39.358	462.7	1.8425
90	29.458	475.5	1.8947		90	37.736	473.1	1.8715
100	28.414	485.7	1.9223		100	36.292	483.5	1.8998
110	27.457	495.9	1.9494		110	34.982	493.9	1.9273
120	26.578	506.3	1.9761		120	33.795	504.4	1.9543
130	25.766	516.7	2.0023		130	32.702	515.0	1.9809
140	25.012	527.2	2.0281		140	31.698	525.6	2.0070

Temp. (°C)		Pressure = 1400kPa			Temp.		Pressure = 1800kPa	
52.39	1086.957	275.7	1.2498	Sat. Liquid	62.87	1041.667	292.6	1.2999
52.39	70.751	424.7	1.7076	Sat. Vapor	62.87	94.520	427.8	1.7022
60	66.577	434.0	1.7357		70	88.188	437.4	1.7306
70	62.228	445.6	1.7700		80	81.517	450.0	1.7667
80	58.715	456.8	1.8023		90	76.373	462.0	1.8001
90	55.775	467.8	1.8331		100	72.172	473.6	1.8317
100	53.241	478.8	1.8628		110	68.652	485.0	1.8618
110	51.032	489.6	1.8915		120	65.606	496.3	1.8909
120	49.056	500.5	1.9194		130	62.920	507.5	1.9191
130	47.283	511.3	1.9467		140	60.530	518.7	1.9466
140	45.674	522.2	1.9734		150	58.390	530.0	1.9734
150	44.198	533.2	1.9997		160	56.437	541.2	1.9997
160	42.839	544.2	2.0255		170	54.641	552.5	2.0256

Temp. (°C)		Pressure = 2000kPa			Temp.		Pressure = 3000kPa	
67.47	1010.101	300.4	1.3223	Sat. Liquid	86.22	877.193	335.3	1.4188
67.47	107.456	428.8	1.6991	Sat. Vapor	86.22	189.568	427.6	1.6758
70	104.329	432.5	1.7101		90	173.927	436.1	1.6992
80	94.819	446.1	1.7493		100	150.440	453.5	1.7466
90	87.957	458.8	1.7845		110	136.372	468.2	1.7855
100	82.577	470.8	1.8173		120	126.267	481.8	1.8204
110	78.165	482.6	1.8483		130	118.400	494.7	1.8528
120	74.432	494.1	1.8781		140	111.972	507.2	1.8835
130	71.198	505.5	1.9068		150	106.546	519.4	1.9128
140	68.340	516.9	1.9347		160	101.860	531.6	1.9411
150	65.794	528.3	1.9619		170	97.740	543.6	1.9686
160	63.497	539.7	1.9884		180	94.070	555.6	1.9953
170	61.401	551.1	2.0145		190	90.765	567.6	2.0215

Abridged and adapted from DuPont Technical Manual T-134a-SI, with permission.

INDEX